DATE DUE

DE ~~2 8~~			
DE ~~19 01~~			

DEMCO 38-296

The Discovery of Subatomic Particles

Staff and students at the Cavendish Laboratory, 1933.
Top row: W. J. Henderson, W. E. Duncanson, P. Wright, G. E. Pringle, H. Miller.
Second row: C. B. O. Mohr, N. Feather, C. W. Gilbert, D. Shoenberg, D. E. Lea, R. Witty, —— Halliday, H. S. W. Massey, E. S. Shire.
Third row: B. B. Kinsey, F. W. Nicoll, G. Occhialini, E. C. Allberry, B. M. Crowther, B. V. Bowden, W. B. Lewis, P. C. Ho, E. T. S. Walton, P. W. Burbidge, F. Bitter.

Fourth row: J. K. Roberts, P. Harteck, R. C. Evans, E. C. Childs, R. A. Smith, G. T. P. Tarrant, L. H. Gray, J. P. Gott, M. L. Oliphant, P. I. Dee, J. L. Pawsey, C. E. Wynn-Williams.

Seated row: —— Sparshott, J. A. Ratcliffe, G. Stead, J. Chadwick, G. F. C. Searle, Professor Sir J. J. Thomson, Professor Lord Rutherford, Professor C. T. R. Wilson, C. D. Ellis, Professor Kapitza, P. M. S. Blackett, —— Davies.

The Discovery of Subatomic Particles

Steven Weinberg

W. H. Freeman and Company
NEW YORK

Library of Congress Cataloging-in-Publication Data

Weinberg, Steven, 1933–
    The discovery of subatomic particles/Steven Weinberg.
        p.   cm.
    Includes bibliographical references.
    ISBN 0-7167-2121-X
    1. Particles (Nuclear physics)   I. Title.
QC793.2.W44   1990   89-24294
539.7'2—dc20   CIP

Printed in the United States of America

1 2 3 4 5 6 7 8 9 0   HA   9 9 8 7 6 5 4 3 2 1 0

To Elizabeth

Contents

Preface

This book grew out of a course that I gave at Harvard in the spring of 1980, as part of the new core curriculum program, and again at the University of Texas as a visitor in 1981. The idea of the course, in brief, was to engage students who were not assumed to have any prior training in mathematics or physics in learning about the great achievements of twentieth-century physics, building up as we went along a background in classical physics—mechanics, electro-magnetism, heat, and so on—as this became necessary in order to understand the more modern developments. I thought the course went well, and had the idea of working up my lecture notes into a textbook, but I have not had the time to carry out this task for all the material of the course. Neil Patterson of W. H. Freeman and Company invited me to present the first part of the story of twentieth-century physics as developed in this course to the readers of *Scientific American* as one of its new Library series, and this book is the result. Perhaps I will be able in future volumes to complete the survey of twentieth-century physics that is begun here.

The book covers the discovery of the fundamental particles that make up all ordinary atoms: the electron, the proton, and the neutron. The general outline is historical, but it is history with one significant difference. Most books about the history of science either are written for readers unfamiliar with the underlying science, and so are forced to be somewhat sketchy and superficial in describing the history, or else they are written for readers already familiar with the science, and so are inaccessible to those who are not. This book is written for readers who may not be familiar with classical physics, but who are willing to pick up enough of it as they go along to be able to understand the rich tangle of ideas and experiments that make up the history of twentieth century physics. This background is provided in a number of "flash-back" sections on the nature of electricity, Newton's laws of motion, electric and magnetic forces, conservation of energy, atomic weights, and so on, which are inserted wherever they are needed to allow the reader to understand the next point in the history.

In fact, I will reveal here (since no one reads prefaces anyway) that these flashback sections and the background material interspersed in other sections represent my secret motivation in writing this book. Like many other scientists, I regard the discoveries of science to be among the most precious elements of twentieth-century culture, and it seems to me a tragedy that so many otherwise well-educated people are cut off from this part of our culture by a lack of familiarity with the basics of science. Yet this educational gap should come as no surprise. Generally speaking, the student or reader who wants to become literate in physics is offered only one path: he or she must follow the same time-honored sequence of courses followed by generations of professional scientists. Always mechanics comes first, followed usually by heat, electricity and magnetism, light, and, as a savory, a little "modern physics." This may be ideal for those students who plan to become physicists, but for many others it seems an impassible desert. Nor is their feeling unreasonable. We physicists are an odd lot, taking great pleasure in the calculations we learn to do in the standard sequence of physics courses: calculations of the collisions of billiard balls, the flow of electric currents in wires, the paths of light rays in a telescope. It just is not reasonable to expect all students or readers to feel this way, any more than we could expect those who never plan to play the piano to enjoy practicing scales. It seems to me that it is this problem of motivation that presents the greatest obstacle when one tries to write for nonscientists about the fundamentals of physics.

My starting point in dealing with this problem was the assumption that, whether or not readers enjoy calculating the collisions of billiard balls, they do generally want to have a cultural background in the revolutionary scientific ideas and discoveries of our own time. Therefore, instead of starting out in this book with a long introduction to elementary classical physics, I invite the reader here to plunge immediately into a sequence of key topics in twentieth-century physics, using each topic as an entrée into just those concepts and methods of classical physics needed to understand that topic. The first topic is the discovery of the first of the elementary particles, the electron. In order to understand the experiments of J. J. Thomson and others that led to this discovery, the reader must learn about Newton's laws of motion, the conservation of energy, and electric and magnetic forces. The next topic is the measurement of the scale of the atom, and here the reader learns more about mechanics and also gets a taste of chemistry. And so on. The point is that the reader is asked to learn aspects of classical physics or chemistry only when it has been made clear that these specific concepts and methods are needed to understand the progress of twentieth-century physics.

It is true that, in a book like this, the order in which principles of elementary physics are introduced cannot be the logical order to which a physicist might be accustomed. For instance, the notion of momentum, which is usually explained at the same time as energy, is not needed here until we come to the discovery of the atomic nucleus; so it is not introduced until then. I do not think that this reordering of topics is necessarily a drawback. Speaking from my own experience, most of what I know about physics and mathematics I have learned only when there was no alternative, when I simply had to learn something in order to get on with my own work. I suspect the same is true of most scientists. So the plan of this sort of book may be closer to the actual education of working scientists than many of the books and courses we design for students who specialize in science.

My hope, then, is that this book may contribute to a radical revision in the way that science is brought to nonscientists. As to whether or not my ideas on how to go about this will work, time and the reader will tell. If all goes well, and I decide to continue this series of books on twentieth-century physics, the next volume will deal with relativity and the quantum theory, and will rely on the foundation of classical physics that is provided here.

This book is intended to be comprehensible to readers who have no prior background in science, and no familiarity with mathematics beyond arithmetic. In the text I include just a few of the most important equations, expressed in words instead of abstract symbols. For those readers who are comfortable with algebra, the Appendixes present some of the calculations that underlie the reasoning described in the main text of the book.

Although this book is written primarily for the nonscientist, it has one aspect that perhaps also my fellow physicists may find interesting. The great scientific achievements described here form a large part of the soil from which our own more recent harvests of discoveries have sprung. Yet I, for one, had only the foggiest idea of the early history of twentieth-century physics when I started to teach the courses at Harvard and Texas, and I suspect that the same is true of many of my colleagues in physics. I hope that scientists may find some of the history (if not the physics) in this book enlightening.

I also hope that this book will be enjoyed by students and practitioners of the history of science, but to them I must make a special apology. It is impossible in a book of this sort to do full justice to the rich tangle of influences that led to the twentieth-century revolutions in physics. All that I could do here was to bring forward a sequence of a few key experimental and theoretical discoveries, which would give me an opportunity to explain elements of classical and modern physics. I have, of course, tried to avoid outright historical errors, but the choice of material and the order of presentation had to be governed by considerations of scientific explanation as well as of history. Certainly I do not intend this book to be regarded as a contribution to historical research. I have, in the course of writing it, read through many of the classic papers of Thomson, Rutherford, Millikan, Moseley, Chadwick, and others, but for the most part I have relied on secondary sources, which are listed in the bibliography at the end of the book. In the Notes at the end of each chapter, I have supplied references to some of the classic articles discussed in the text, and to more recent works on which I specially relied.

I am very grateful to Howard Boyer, Andrew Kudlacik, Neil Patterson, and Gerard Piel for their friendly cooperation in readying this book for publication. My gratitude is also due to Aidan Kelly for his sensitive editing and many helpful suggestions. Paul Bamberg provided valuable assistance when at Harvard I first taught this course. For taking the trouble to read and comment on various portions of the book, I wish to express warm thanks to I. Bernard Cohen, Peter Galison, Gerald Holton, Arthur Miller, and Brian Pippard. Many grievous historical errors were avoided through their help.

Steven Weinberg
Austin, Texas
May 1982

Introduction and Update

In the countryside south of Dallas construction will soon begin on a new scientific instrument. It is variously known as the Superconducting Supercollider, or SSC, or just the Supercollider. The scale of this construction is extraordinary. If all goes well and funding is continued, a 53-mile-long oval tunnel will be dug through the Austin Chalk and Taylor Marl that underlie the farms and little towns of Ellis County, Texas. Running through this tunnel will be a pair of vacuum pipes. When the Supercollider is operating, each pipe will carry a beam of protons, electrically charged particles that are found in all atomic nuclei. As the two beams of protons go in opposite directions round and round the oval track, electromagnetic waves will gradually accelerate the protons to the unprecedentedly high energy of twenty trillion electron volts. Thousands of powerful magnets surrounding the vacuum pipes will focus the two beams of protons and curve them so that they follow the correct paths. At several stations within the tunnel the beams will be made to collide head-on, and enormous devices, some the size of battleships, will sort out the showers of particles produced in the collisions. Teams of experimental physicists will study the data that is produced, searching for signs of new forms of matter, new forces, new clues to the laws that govern all physical phenomena.

We can anticipate that when the Supercollider is ready to begin operations it will become for a while a center of attention for the world's press. Anyone with any interest in the physical sciences will want to know the latest news from Ellis County. It would be a pity if this were not the case. The enterprise of science is one of the brightest parts of the culture of the twentieth century, one that (as at least we scientists feel) ought to attract the attention of every educated person. Also, United States taxpayers will have paid most of the cost of the Supercollider, and might as well get their full money's worth. Still, there is something disturbing about such a sudden focusing of public attention on a new scientific instrument. I recall that Richard Feynman once said something to the effect that he did not understand why many of the public always wanted to know the latest news about science, but had little interest in learning about what went before it. After all, the story of scientific progress is above all a *story,* and it helps in reading the later chapters of a story to know something about what happened in the earlier chapters.

The Supercollider will in fact be opening one more chapter in a story that has been going on for some time. It has been designed to fill in the gaps in a physical theory that we now call the Standard Model of Elementary Particles and, more importantly, to provide clues that may lead to a deeper theory underlying the Standard Model. The Standard Model is not today's news; it was put into essentially its present form through the work of a number of theorists in the late 1960's and early 1970's, was confirmed by experiments through the 1970's, and has roots extending far back in the physics of this century.

This is not the place for a thorough introduction to the Standard Model, which is not the subject of this book, but a brief description is useful. The Standard Model is mathematically expressed as a theory of the type known as a quantum field theory, in which elementary particles of various types appear as bundles of the energy and momentum of corresponding types of fields and all the forces between the particles are produced by the exchange of other elementary particles. The whole theory is governed by certain principles of symmetry, which require that the equations describing the fields remain invariant if one makes various changes in one's frame of reference, and which give the theory much of its predictive power.

According to the Standard Model, the elementary particles form five broad classes. There are the leptons, including the electrons that make up the outer parts of all ordinary atoms, together with other particles that are more or less like electrons but much heavier and unstable, and others, the neutrinos, which apparently have no mass at all. There are the quarks, of which two types, known as the "up" and "down" quarks, make up the protons and neutrons, the particles inside the nuclei of ordinary atoms. The other quark types are unstable and much heavier. (We shall have more to say about these quarks and leptons in the final chapter of this book.) There are the vector bosons (awkward name, that), particles more or less like the particle of light, the photon. Some, however, the "W", and "Z" particles, are very heavy and unstable, and others, the gluons, though massless, share with the quarks the weird property of being impossible to isolate from other quarks or gluons. The various strong, weak, and electromagnetic forces among quarks and leptons are produced by the exchange of these vector bosons. Standing in a fourth class all its own is the graviton, the massless particle of gravitational radiation, which does not seem to have much to do with the other particles of the Standard Model but which is needed to explain the gravitational forces acting among all particles. Finally, there is a mysterious fifth class of particles, whose interactions are believed to be responsible for producing the masses of all of the particles of the Standard Model but whose properties are still largely unknown. (The much-discussed but so far unobserved Higgs boson would be a member of this class.) The target value of twenty trillion volts planned for the energy of the Supercollider was chosen in part because it is at such energies that effects of the interactions of this fifth class of particles are sure to appear.

But the Supercollider does not only aim at clearing up the mystery of the fifth class of particles and their interactions. Even with all that understood, the Standard Model will obviously not be the final expression of the fundamental laws of physics. For one thing, we will still have to understand why the Standard Model is the way it is. Why just so many quarks and leptons? Why does the model obey symmetry principles that dictate that there must be just eight types of gluons along with one W, one Z, and one photon? What does the graviton have to do with all this? And why do all the constants of the theory, the masses, charges, and such, take the numerical values they do?

Over the past decade theoretical physicists have struggled with a variety of more or less speculative ideas, which it was hoped would help us to see the deeper, simpler theory that we feel must underlie the Standard Model. Theories involving new symmetry principles, higher dimensions, strings instead of point particles, and even fluctuations in the connectedness of space and time have been explored mathematically. I think that this work will provide much of the intellectual capital on which we shall have to draw in the decades to come. In particular, the string theories have at last provided a mathematical framework for describing gravitons in the same quantum-mechanical terms as other particles. But it must be admitted that a decade of brilliant mathematical speculation has not yet produced anything new in the way of precise numerical predictions, verified by experiment, that could convince us that we are on the right track. This of course is why elementary-particle physicists feel that it is so important for them to get their hands on new data of the sort that can only be provided by new experimental facilities like the Supercollider.

With all its holes and loose ends, the Standard Model still represents quite an impressive unification and simplification of our understanding of elementary particles, one that was reached only after decades of experimental and theoretical effort. This book deals mostly with the earlier parts of the story. Still, the *dramatis personae* that we shall encounter in these beginnings are the same that will appear, though perhaps in a transformed role, in the Standard Model. The proton, whose existence was inferred by Ernest Rutherford from experiments performed under his leadership in 1911, and the neutron, discovered by James Chadwick in 1932, are not only the cages in which the "up" and "down" quarks of the Standard Model come imprisoned within atomic nuclei; they are also analogs for these quarks, because they share the same sort of two-member nuclear family relationship. The electron, discovered by J. J. Thomson in 1897, is the prototype of all the leptons of the Standard Model, and continues as a paradigm for all of the elementary particles, stubbornly refusing to reveal any hints of being constructed of more fundamental constituents.

This book aims not only at describing the first appearances in physics of particles like the electron, the proton, and the neutron but also at introducing the reader to certain themes that occur again and again in today's theories and experiments. The language of physics continues to deal with concepts like energy, electric charge, electric and magnetic field, mass, momentum, and so on. Most people today have at least a rough idea of what these quantities signify, but it is a great help in understanding today's scientific discoveries to pin their meaning down more precisely. For this purpose one can hardly do better than to see how these concepts appear in the historic experiments of Thomson, Rutherford, Chadwick, and their contemporaries.

Then there are concepts that are a little more esoteric, like the idea of an effective cross section or a mean lifetime. We will see here that Rutherford's discovery of the atomic nucleus was based on the measurement in his laboratory of the effective cross section for the scattering of alpha particles (the nuclei of helium) by atoms of gold, and that the first measurement of a mean lifetime was also made by Rutherford in his study of the radioactive gas then known as thorium emanation. A good deal of the work done at high-energy laboratories like the Supercollider consists of measurements of cross sections and mean lifetimes of one sort or another.

Even the idea of an accelerator goes back to the beginnings of elementary particle physics. As we shall see, Thomson's apparatus was itself a particle accelerator, although its energy, which was supplied by ordinary storage batteries instead of electromagnetic waves, was only a few thousand volts instead of twenty trillion volts, and Thomson's interest was with the beam itself rather than its collisions.

The Supercollider is now scheduled to begin operations in mid-1998, but it would be wonderfully appropriate if a slight speedup allowed operations to begin in 1997, just in time for the centennial of the discovery of the electron, the first known of all the elementary particles.

•   •   •

This new edition of *The Discovery of Subatomic Particles* gives me a chance to bring several items of the earlier edition up to date.

xviii   INTRODUCTION AND UPDATE

• 	In Chapter 4, "The Nucleus," I reported several explanations that I had heard for the decision by Peter Kapitsa to have a crocodile carved on a wall of a Cambridge laboratory as a symbol of Rutherford. Since then, I have received an interesting letter from Mr. J. L. Koffman of Leicester, England, who was acquainted with Kapitsa's three elder half sisters. According to Koffman, there was a "rather rude song" about a crocodile, that was popular in Russia in the years 1922 to 1925. In this song, a large crocodile walks about the streets grabbing persons of various nationalities by more and more intimate parts of their bodies. I suppose that the idea must have been to use the crocodile as a symbol of Rutherford's ferocity in a generalized sense and not as a comment on his particular habits. In any case, I suspect that I have not yet heard the last explanation of the Rutherford crocodile.

• 	Professor George Kauffman of California State University in Fresno has written to let me know that several of the advances in nuclear physics described in this book were actually pioneered by the well-known physical chemist William Draper Harkins (1873–1951). I find this confirmed in an essay by T. F. Young, who credits Harkins with a number of early advances, including the first calculation of the energy that might be produced in the sun by the conversion of hydrogen to helium and the use of neutrons in scientific experiments immediately after their discovery by Chadwick.

• 	In Chapter 5, "More Particles," I said that cosmic neutrinos have never been observed. This is still true for the huge number of low-energy neutrinos left over from the first few minutes of the big bang, but in January 1987 neutrinos from outside the solar system were detected in several underground laboratories. These neutrinos were emitted in the great supernova explosion, SN 1987A, in the Larger Magellanic Cloud, some 150,000 light years from earth. The fact that after this enormous voyage the neutrinos all arrived within a few seconds of each other shows that despite their differing energies, they all had essentially the same velocity, as would be expected if they were massless particles that (like photons) always travel at the speed of light. This, together with improvements in laboratory experiments on neutrinos from the radioactive hydrogen isotope tritium, shows that the neutrino mass cannot be greater than about ten electron volts, or one fifty-thousandth of the electron mass.

• 	In the same chapter, I mentioned that in 1982, at the time of writing, physicists were awaiting the discovery of one more type of quark, as well as particles called intermediate vector bosons (another name for the W and Z particles mentioned above) and Higgs bosons. Since then, the W and Z particles have been discovered in the collisions of high-energy protons and antiprotons at the CERN laboratory near Geneva, Switzerland. It is very gratifying that these particles turn out to have just the masses predicted by the Standard Model: the W and Z masses are 86.4 and 98.5 proton masses, respectively. The missing quark type (the "top" quark) and the Higgs boson continue to elude detection.

•   The biographical information given in the previous edition of this book for some of the physicists who played a leading role in the history of elementary particle research must sadly now be completed as follows:

Peter Leonidovitch Kapitsa (1894–1984)
Paul Adrien Maurice Dirac (1902–1984)
Emilio Gino Segré (1905–1989)

We will miss their genius in the work left before us.

Steven Weinberg
Austin, Texas
October 1989

The Discovery of Subatomic Particles

1

A World of Particles

1

A World of Particles

How many men and women, studying the tiny boulders in a handful of sand, may have conceived of the finer and harder grains that make up all forms of matter? The explicit statement that matter is composed of indivisible particles called atoms (from the Greek ατομοσ, "uncuttable") we trace to the ancient town of Abdera, on the seacoast of Thrace. There, in the latter part of the fifth century B.C., the Greek philosophers Leucippus and Democritus taught that all matter is made up of atoms and empty space.

Abdera now lies in ruins. No word written by Leucippus has survived, and of the writings of Democritus we have only a few unhelpful fragments. However, their idea of atoms survived and was quoted endlessly in the following millennia. This idea allows one to make sense of a great many commonplace observations that would be quite puzzling if it were thought that matter was a continuum that fills the space it occupies. How better can we understand the dissolving of a piece of salt in a pot of water than by supposing that the atoms of which the salt is composed spread into the empty spaces between the atoms of the water? How better can we understand the spread of a drop of oil on the surface of water, out to a definite area and no farther, than by supposing that the film of oil spreads until it is a few atoms thick?

After the birth of modern science, the idea of atoms came to be used as a basis of quantitative theories of matter. In the seventeenth century Isaac Newton (1642–1727) attempted to account for the expansion of gases in terms of the outrush of their atoms into empty space. More influentially, in the early nineteenth century John Dalton (1766–1844) explained the fixed ratios of the weights of chemical elements that make up common compounds in terms of the relative weights of the atoms of these elements.

By the end of the nineteenth century the idea of the atom had become familiar to most scientists—familiar, but not yet universally accepted. Partly because of the heritage of Newton and Dalton, there was a disposition to use atomic theories in England. On the other hand, resistance to atomism persisted in Germany. It was not so much that the German physicists and chemists positively disbelieved in atoms. Rather, under the influence of an empiricist philosophical school centered on Ernst Mach (1836–1916) of Vienna, many of them held back from incorporating into their theories anything that—like atoms—could not be observed directly. Others, like the great theorist Ludwig Boltzmann (1844–1906), did use atomistic assumptions to build theories of phenomena such as heat, but had to suffer the disapproval of their colleagues; it is said that the opposition to Boltzmann's work by the followers of Mach contributed to Boltzmann's suicide in 1906.

All this changed in the first decades of the twentieth century. Oddly, the general acceptance of the atomic nature of matter came about through the discoveries of the constituents of the atom, the electron and atomic nucleus— discoveries that undercut the old idea that atoms are indivisible. These discoveries are the subject of this book. Before we go into the history of these discoveries, however, let us anticipate them and recall what is now understood about the constituents of the atom. This is only a brief overview; we will be going into all this in much greater detail in the subsequent chapters of this book.

Most of the mass of any atom is contained in the small, dense nucleus at its center, which carries a positive electric charge. Revolving in orbit around the nucleus are one or more electrons, which carry negative electric charge and are held in orbit by the force of electrical attraction. The typical radius of an electron's orbit is about 10^{-10} meters* (a unit of length called the angstrom), while the nucleus is much smaller, with typical diameter of about 10^{-15} meters (a unit called the fermi). The various chemical elements each consist of atoms of one specific kind, the atoms of one element differing from those of another element in the number of electrons they contain: one for hydrogen, two for helium, and so on up to 103 for lawrencium. Atoms can combine into larger aggregates—molecules—by lending, trading, or sharing their electrons; each chemical compound consists of molecules of one specific kind. Under ordinary circumstances, visible light is absorbed or emitted when the electrons in an atom or molecule are excited to orbits of higher energy or sink back to orbits of lower energy, respectively. Electrons can also be shaken loose from atoms, and by traveling through a metal wire produce an ordinary electric current.

In all these phenomena—chemical, optical, and electrical—the nucleus of the atom is essentially inert. However, the nucleus itself is a composite system with its own constituents, the particles known as protons and neutrons. The proton carries an electric charge equal and opposite to that of the electron; the neutron is electrically neutral. The proton has a mass of 1.6726×10^{-27}

---

* For a short discussion of scientific notation, see the box at the end of this chapter.

kilograms, the neutron's mass is a little larger (1.6750×10^{-27} kilograms), and the electron's mass is much smaller (9.1095×10^{-31} kilograms). The protons and neutrons in nuclei, like the electrons surrounding the nuclei, can be excited to states of higher energy or, if excited, can fall back to a state of lower energy, but the energies needed to excite the nuclear particles in the nucleus are typically a million times those needed to excite the electrons in the outer part of the atom.

All ordinary matter is composed of atoms, which in turn consist of protons, neutrons, and electrons. However, it would be a mistake to conclude that protrons, neutrons, and electrons make up the whole list of fundamental entities. The electron is just one member of a family of particles called leptons, of which some half dozen are now known. The proton and the neutron are members of a much larger family of particles called hadrons, of which hundreds are known. The special property that makes electrons, protons, and neutrons the ubiquitous ingredients of ordinary matter is their relative stability. Electrons are believed to be absolutely stable, and protons and neutrons (when bound in an atomic nucleus) live at least 10^{30} years. With a few exceptions, all other particles have very short lifetimes, and are therefore very rare in the present universe. (The only other stable particles are those that have zero or very small values of mass and charge and therefore cannot be trapped into atoms or molecules.)

The proton, the neutron, and the other hadrons are now believed to be composites themselves, made up of more elementary constituents called quarks. As far as is known, the electron and the members of the lepton family are truly elementary. But elementary or not, it is the particles that make up ordinary atoms—protons, neutrons, and electrons—that will concern us in this book.

Just as ancient Abdera symbolizes for us the birth of atomism, there is one place with which the discovery of the constituents of the atom is especially associated: It is the Cavendish Laboratory of the University of Cambridge. There, in 1897, Joseph John Thomson (1856–1940) performed the experiments on cathode rays that led him to conclude that there is a particle—the electron—that is both the carrier of electricity and a basic constituent of all atoms. It was at the Cavendish in 1895–1898 that Ernest Rutherford (1871–1937) began his work on radioactivity, and to the Cavendish in 1919 that Rutherford returned, after his discovery of the atomic nucleus, to succeed Thomson as Cavendish Professor of Experimental Physics and to found what was long the preeminent center for nuclear physics. The list of constituents of the atom was completed at the Cavendish in 1932, when James Chadwick (1891–1974) discovered the neutron.

The exterior of the Cavendish Laboratory at Cambridge as it appeared from Maxwell's time onward. The building is now used for other university purposes, and the laboratory has been moved to more modern quarters.

I first visited the Cavendish Laboratory in the spring of 1962, when as a very junior physicist I was on leave from the University of California at Berkeley for a year in London. The laboratory then still occupied its original gray stone buildings in Free School Lane, where it had stood since 1874, on land purchased by the University of Cambridge in 1786 for use as a botanical garden. I remember it as a warren of little rooms connected by an incomprehensible network of stairways and corridors. It was very different from California's great Radiation Laboratory, which looked commandingly out over the bay from its sunlit site in the Berkeley hills. The Cavendish Laboratory gave the impression that it was the scene not so much of a massive assault on the secrets of nature, but of a guerilla campaign, an effort of limited resources, in which the chief weapons were the cleverness and bravado of gifted individuals.

The Cavendish Laboratory had its origin in the report of a university committee that met in the winter of 1868–69 to consider how to make a place for experimental physics in Cambridge. It was a time of widespread enthusiasm for experimental science. A great new laboratory for experimental physics had recently been opened in Berlin, and university laboratories were being constructed at Oxford and Manchester. Cambridge had not played a leading role in experimental science, despite (or possibly because of) a tradition of excellence in mathematics that went back to the seventeenth-century Lucasian Professor of Mathematics, Sir Isaac Newton. But empiricism was now in the air, and the committee called for a new Professorship of Experimental Physics and a new building to house lectures and experiments.

It remained to find funds and a professor. The first need was quickly met. The chancellor of the university at that time was William Cavendish, seventh Duke of Devonshire and a member of the family that had earlier produced the distinguished physicist Henry Cavendish (1731–1810), who first measured the attractive force of gravity between laboratory masses. Devonshire had done spectacularly well in mathematics as an undergraduate at Cambridge, and had then gone on to do even better making money in the Lancashire steel industry. In October 1870 he wrote to the vice-chancellor of the university, offering to provide the funds required for building and apparatus—some £6,300. When the building was completed in 1874, a letter of thanks (in Latin) was presented to Devonshire, proposing to name the laboratory after the Cavendish family.

It was hoped that the first Cavendish Professor would be Sir William Thomson (1824–1907), later Lord Kelvin, the most eminent experimental physicist in Britain. However, Thomson wished to stay in Glasgow, and instead the Cavendish Professorship went to another Scot: James Clerk Maxwell

(1831–1879), who at age 39 was living in retirement on his estate at Glenair.

Maxwell is generally regarded as the greatest physicist between Newton and Einstein, but it is odd to think of him as a professor of experimental physics. Although he did experimental work of some distinction on color perception (with his wife as colleague) and on electrical resistance, his greatness rests almost entirely on his theoretical work. Above all, it was Maxwell who completed the equations that describe the phenomena of electricity and magnetism, and then used these equations to predict the existence of electromagnetic waves, thereby explaining the nature of light. Although Maxwell's work lent great prestige to the Cavendish Professorship, the Cavendish Laboratory did not develop into a leading center of experimental physics during his tenure. For instance, the existence of electromagnetic waves was demonstrated experimentally not at the Cavendish Laboratory, but rather at Karlsruhe, by the German experimentalist Heinrich Hertz (1857–1894).

After Maxwell's death in 1879, the Cavendish Professorship was again offered to William Thomson, who again declined. This time the professorship went to John William Strutt (1842–1919), the third Lord Rayleigh. Rayleigh, who was gifted both as a theorist (though not on Maxwell's level) and as an experimentalist, worked on a huge variety of physical problems. Even today, when one is confronted with a problem in hydrodynamics or optics, a good place to start in looking for a solution is in his collected works. Under Rayleigh's leadership the Cavendish Laboratory remained small, most of its research consisting of Rayleigh's own work, but important improvements were made. New apparatus was purchased, instruction was reorganized, a workshop was opened, and beginning in 1882 women were admitted on the same terms as men. In 1884 Rayleigh resigned the Cavendish Professorship, and shortly thereafter he accepted the less demanding position of professor at the Royal Institution in London.

Once again, the Cavendish Professorship was offered to William Thomson (now Lord Kelvin), and once again Kelvin decided to stay in Glasgow. The obvious next choice was between R. T. Glazebrook and W. N. Shaw, who did most of the work of preparing apparatus for lectures and experiments. To almost everyone's surprise, the professorship went instead to a young man of mostly mathematical talent—J. J. Thomson. Although it is not clear if there were any good reasons then for this decision, it was the right choice. Following Rayleigh's advice, Thomson began his epoch-making experimental work on cathode rays. What is more, under his direction the Cavendish Laboratory came alive. Flocks of talented experimentalists came to work there, including in 1895 a young New Zealander, Ernest Rutherford. The stage was now set for the discovery of the constituents of the atom.

Scientific or Exponential Notation

Atoms and subatomic particles are very small, and there are a large number of them in any ordinary piece of matter. We can get nowhere in speaking of them without using the convenient "scientific" or "exponential" notation for very large and very small numbers. This notation employs powers of ten: 10^1 is just 10; 10^2 is the product of two tens, or 100; and so on. Also, 10^{-1} is the reciprocal of 10^1, or 0.1; 10^{-2} is the reciprocal of 10^2, or 0.01, and so on. (That is, 10^n is a one with n zeroes, and 10^{-n} is a decimal point followed by $n - 1$ zeroes and a one.) Here is a list of some powers of 10, with their American names and the prefixes that are used to denote them.

Power of 10	American name	Prefix
10^1	ten	deka
10^2	hundred	hecto
10^3	thousand	kilo
10^6	million	mega
10^9	billion	giga
10^{12}	trillion	tera
10^{-1}	tenth	deci
10^{-2}	hundredth	centi
10^{-3}	thousandth	milli
10^{-6}	millionth	micro
10^{-9}	billionth	nano
10^{-12}	trillionth	pico
10^{-15}	quadrillionth	femto

(I say American here because for the British a billion is 10^{12}, and 10^9 is a milliard.) For instance, 10^3 grams is one kilogram; 10^{-2} meter is one centimeter; and 10^{-3} ampere is one milliampere. The great thing about this scientific notation is not just that it saves writing words like "quadrillionth," but that it makes arithmetic so easy. If we want to multiply 10^{23} times 10^5, we are taking the product of 23 tens times the product of 5 tens, or 28 tens in all; so the answer is 10^{28}. Likewise if we want to multiply 10^{23} by 10^{-19} (or divide 10^{23} by 10^{19}), then we are dividing the product of 23 tens by the product of 19 tens; so the answer is 10^4.

This is the general rule: in multiplying powers of 10 add the powers; in dividing subtract them. According to this rule ten to any power divided by ten to the same power would be ten to the zeroth power; so 10^0 is taken to be 1. To deal with numbers that are not simply a power of 10, we can always write them as a number between 1 and 10 times a power of 10: thus 186,324 is 1.86324×10^5, and

0.0005495 is 5.495×10^{-4}. In multiplying or dividing such numbers, we multiply or divide the numbers that accompany the powers of ten and combine the powers of ten as before; thus 1.86324×10^{5} times 5.495×10^{-4} is 1.86324 times 5.495, or 10.238, times $10^{5} \times 10^{-4}$, or 10^{1}, which we could also write as 1.0238×10^{2}. These days scientific notation is used pretty widely, from the pages of *Scientific American* to the electronic calculators that one can buy for under \$20 (2×10^{1} dollars), so I will use it freely in this book.

2

The Discovery
of the Electron

2

The Discovery of the Electron

This century has seen the gradual realization that all matter is composed of a few types of elementary particles—tiny units that apparently cannot be subdivided further. The list of elementary particle types has changed many times during the century, as new particles have been discovered and old ones have been found to be composed of more elementary constituents. At latest count, there are some sixteen known types of elementary particles. But through all these changes, one particle type has always remained on the list: the electron.

The electron was the first of the elementary particles to be clearly identified. It is also by far the lightest of the elementary particles (aside from a few types of electrically neutral particles that appear to have no mass at all) and one of the few that does not decay into other particles. As a consequence of its lightness, charge, and stability, the electron has a unique importance to physics, chemistry, and biology. An electrical current in a wire is nothing but a flow of electrons. Electrons participate in the nuclear reactions that produce the heat of the sun. Even more important, every normal atom in the universe consists of a dense core, (the nucleus) surrounded by a cloud of electrons. The chemical differences between one element and another depend almost entirely on the number of electrons in the atom, and the chemical forces that hold atoms together in all substances are due to the attraction of the electrons in each atom for the nuclei of the other atoms.

The discovery of the electron is usually and justly credited to the English physicist Sir Joseph John Thomson (1856–1940). Thomson went up to the University of Cambridge as a scholarship student in 1876. After placing second in the competitive mathematical "tripos" examination in 1880, he earned a fellowship at Trinity, the old Cambridge college of Isaac Newton, and remained a fellow of Trinity for the following 60 years of his life. Thomson's early work was chiefly mathematical, and not outstandingly important; so he was somewhat surprised when in 1884 he was elected to the Cavendish Professorship of Experimental Physics. It was in his experimental researches and his leadership of the Cavendish Laboratory from 1884 to 1919 that Thomson

J. J. Thomson.

made his greatest contributions to physics. He was actually not skillful in the execution of experiments; one of his early assistants recalled that "J.J. was very awkward with his fingers, and I found it necessary not to encourage him to handle the instruments." His talent—one that is for both theorists and experimentalists the most important—lay instead in knowing at every moment what was the next problem to be attacked.

From what is written about him, I gather that Thomson was greatly loved by his colleagues and students. It is certain that he was greatly honored: by the Nobel Prize in 1906, a knighthood in 1908, and the Presidency of the Royal Society in 1915. He served Britain in World War I as a member of the Board of Investigation and Research, and in 1918 was appointed Master of Trinity College, a post he held until shortly before his death. He was buried in Westminster Abbey, not far from Newton and Rutherford.

Shortly after assuming the Cavendish Professorship, Thomson began his investigation of the nature of discharges of electricity in rarefied gases, and in particular the type of discharge known as cathode rays. These spectacular phenomena were interesting enough in themselves, but their study led Thomson to an even more interesting problem: that of the nature of electricity itself.

His conclusion, that electricity is a flow of the particles that are today known as electrons, was published in three papers in 1897.[1] But before we take up Thomson's investigations, let us review earlier efforts to understand the nature of electricity.

Flashback: The Nature of Electricity*

It has been known since early times that a piece of amber, when rubbed with fur, will acquire the power to attract small bits of hair and other materials. Plato refers in his dialogue *Timaeus* to the "marvels concerning the attraction of amber."[2] By the early Middle Ages, it had become known that this power is shared by other materials, such as the compressed form of coal known as jet. The earliest written observation of this property of jet seems to be that of the Venerable Bede (673–735), the English monk who also studied the tides, calculated the dates of Easter for centuries to come, and wrote one of the world's great works of history, *The Ecclesiastical History of the English*. In his history, Bede notes of jet that "like amber, when it is warmed by friction, it clings to whatever is applied to it."[3] (Bede exhibits here a confusion about the cause of electric attraction, between friction itself and the warmth it produces—a confusion that was often to recur until the eighteenth century.) Other substances, such as glass, sulfur, wax, and gems, were found to have similar properties by the English physician William Gilbert (1544–1603), president of the Royal College of Surgeons and court physician to Elizabeth I and James I. It was Gilbert who introduced the term *electric* (in his Latin text, *electrica*), after the Greek word *electron* (ηλεκτρον) for amber.[4]

The observation of electrical attraction in so many different substances led naturally to the idea that electricity is not an intrinsic property of the substances themselves, but is instead some sort of fluid (to Gilbert, an "effluvium") that is produced or transferred when bodies are rubbed together and spreads out to draw in nearby objects. This picture was supported by the discovery by Stephen Gray (1667–1736) of electrical conduction. In 1729, while a "poor brother" of the Charterhouse in London, Gray reported in a letter to some fellows of the Royal Society that "the Electrick Virtue" of a rubbed glass tube may be transmitted to other bodies, either by direct contact or via a thread connecting them, so "as to give them the same Property of attracting and repelling light Bodies as the Tube does."[5] It was clear that,

---

*This is an often-told story, and my recount of it here is based almost entirely on secondary sources. I review it here because it gives a good idea of what was known and what was not known about electricity when the experiments on cathode rays began.

whatever electricity might be, it could be separated from the body in which it was produced. But the problem of the nature of electricity became more complicated when it was found that electrified bodies could either attract or repel other electrified bodies, raising the question whether there was one kind of electricity or two.

Among those who first observed electric repulsion were Niccolo Cabeo (1586–1650)[6] and Francis Hauksbee (1666–1713), a paid demonstrator of scientific experiments at the Royal Society of London. In a communication to the Royal Society in 1706 Hauksbee reported that, when a glass tube was electrified by rubbing, it would at first attract bits of brass leaf, but that after the bits of brass came in contact with the tube they would be repelled by it.

Further complications were discovered in France by one of the most versatile scientists of the eighteenth century, Charles-François de Cisternay Du Fay (1698–1739). Chemist at the Académie des Sciences and administrator of the Jardin Royal des Plantes, Du Fay wrote papers on almost every conceivable scientific subject, including geometry, fire pumps, artificial gems, phosphorescence, slaked lime, plants, and dew. In 1733 he learned of Stephen Gray's experiments and began to work on electricity. Soon he observed that bits of metal that had been in contact with an electrified glass tube would repel each other (as observed by Cabeo and Hauksbee) but would *attract* bits of metal that had been in contact with an electrified piece of a resin, copal. Du Fay concluded that "there are two electricities, very different from each other; one of these I call vitreous electricity; the other resinous electricity."[7] "Vitreous" electricity (from the Latin *vitreus,* glassy) is produced when substances like glass, crystal, or gems are rubbed, especially with silk. "Resinous" electricity is produced when resins like amber or copal are rubbed, especially with fur. At the same time, the silk used to rub the glass picks up resinous electricity, and the fur used to rub the resin picks up vitreous electricity. Both vitreous and resinous electricity were assumed to attract ordinary matter, and vitreous electricity was assumed to attract resinous electricity, but bodies carrying vitreous electricity were assumed to repel each other, and likewise for resinous electricity. That is, unlike types of electricity attract each other, but like types repel. A bit of metal that had come into contact with the rubbed glass tube would pick up some of the tube's vitreous electricity, and would therefore be repelled by it; and a bit of metal that had been in contact with a rubbed amber or copal rod would pick up some of the rod's resinous electricity, and so again would be repelled by it, but the two bits of metal would attract each other, because they would be carrying electricity of two different types.

Gray and Du Fay did not write of electricity as a fluid, but rather as a condition that could be induced in matter. It was the Abbé Jean-Antoine Nollet

(1700–1770), preceptor to the French royal family and professor at the University of Paris, who interpreted Du Fay's two types of electricity specifically as two distinct types of electrical fluid, one vitreous and the other resinous.

The two-fluid theory was consistent with all experiments that could be carried out in the eighteenth century. But physicists' passion for simplicity does not let them rest with a complicated theory when a simpler one can be found. The two-fluid theory of electricity was soon to be challenged by a one-fluid theory, proposed first by the London physician and naturalist William Watson (1715–1787) and then more comprehensively and influentially by the Philadelphia savant Benjamin Franklin (1706–1790).

Franklin became interested in electricity when in 1743, on a visit to Boston, he happened to witness electrical experiments carried out by a Dr. Adam Spencer, a popular lecturer from Scotland. Soon Franklin received some glass tubes and instructions from a correspondent in London, the manufacturer and naturalist Peter Collinson, and began his own experiments and speculations, which he reported in a series of letters to Collinson. In brief, Franklin concluded that electricity consisted of a single kind of fluid, consisting of "extremely subtle particles," which could be identified with what Du Fay had called vitreous electricity. (Franklin did not know of Du Fay's work, and did not use his terminology.) Franklin supposed ordinary matter to hold electricity like a "kind of spunge." When a glass tube is rubbed with a silk cloth, some of the electricity from the silk is transferred to the glass, leaving a deficiency in the silk. It is this deficiency of electricity that is to be identified with what Du Fay called resinous electricity. Similarly, when an amber rod is rubbed with fur, some electricity is transferred, but this time from the rod to the fur, leaving a deficiency of electricity in the rod; again, the deficiency of electricity in the rod and the excess in the fur are to be identified with Du Fay's resinous and vitreous electricity, respectively. Franklin referred to a deficiency of electricity as *negative* electricity and to an excess as *positive* electricity; the amount of electricity (positive or negative) in any body he called the electric *charge* of the body. These terms are the ones that are still in general use today.

Franklin also introduced the fundamental hypothesis of the conservation of charge. Electricity is never created or destroyed, but only transferred. Hence, when a glass rod is rubbed with silk, the positive electric charge on the rod is exactly equal numerically to the negative charge on the silk; balancing positive and negative, the total charge remains zero.

What about attraction and repulsion? Franklin supposed that electricity repels itself but attracts the matter that holds it. Thus, the repulsion that Cabeo observed between pieces of brass leaf that had been in contact with a rubbed glass rod could be understood because these bits of metal all contained an

Benjamin Franklin in 1762. Notice the apparatus behind him; the position of the two balls indicates that a charged cloud is overhead.

excess of electricity, while the attraction that Du Fay observed between such bits of metal and others that had been in contact with a rubbed rod of resin could be understood because the latter bits had a deficiency of electricity, so that the attraction between their matter and the former bits' electricity would dominate. This neatly accounted for the repulsion observed between two bodies each carrying the "vitreous" electricity, and for the attraction observed between a body carrying "resinous" electricity and one carrying "vitreous" electricity.

But then what about the repulsion between two bodies carrying resinous electricity, such as bits of metal that had been in contact with a rubbed amber rod? This gap in Franklin's one-fluid theory was filled by Franz Ulrich Theodosius Aepinus (1724–1802), director of the astronomical observatory in St. Petersburg. After learning of Franklin's ideas, Aepinus in 1759 suggested that, in the absence of a counterbalancing quantity of electricity, ordinary matter repels itself.[8] Thus, the repulsion between bodies that had been supposed to carry resinous electricity was explained in terms of the repulsion between the matter of the bodies when it was stripped of some of its normal accompaniment of electricity. With this emendation, the one-fluid theory of Franklin was thus able to account for all the phenomena that had been explained by the two-fluid theory of Du Fay and Nollet.

Franklin's letters were assembled by Collinson into a book, which by 1776 had gone through ten editions, some in English and others in Italian, German, and French.[9] Franklin became a celebrity; he was elected to the Royal Society of London and the French Académie des Sciences, and his work influenced all later studies of electricity in the eighteenth century. Indeed, Franklin's fame was a great asset to the thirteen American colonies when, during the revolutionary war, Franklin served as the American minister to France. However, despite Franklin's enormous prestige, the question of one fluid or two continued to divide physicists until well into the nineteenth century, and it was only really settled with the discovery of the electron.

For readers who cannot wait until we come to the discovery of the electron to learn whether the one-fluid or the two-fluid theory is correct—the answer is that they were both correct. Under normal circumstances, electricity is carried by the particles called electrons, which as Franklin supposed possess electricity of only one type. But Franklin guessed wrong as to which type of electricity it was. In fact, electrons carry electricity of the type that Du Fay had called "resinous," not the "vitreous" type. (Physicists continue to follow Franklin's lead in calling vitreous electricity positive and resinous electricity negative, so we are stuck in the unfortunate position of saying that the most common carriers of electricity carry negative electrical charge.) Thus, when a glass tube is rubbed with silk, the tube picks up vitreous electricity and the silk acquires resinous electricity, because electrons are transferred from the tube to the silk. On the other hand, when an amber rod is rubbed with fur, electrons are transferred from the fur to the rod.

In the atoms of ordinary matter, electrons are bound to dense atomic nuclei, which contain most of the mass of any substance and are normally immobile in solids. As Franklin supposed, electrons repel electrons, and electrons and nuclei attract each other; and as Aepinus supposed, atomic nuclei repel other nuclei. But it is convenient to think of the positive or vitreous charge of matter as residing in the nuclei, and not as merely an absence of electrons. Indeed, by dissolving solids like salt in water it is possible to shake the atomic nuclei loose (though they will almost always be accompanied by some electrons), and in this case it is possible to have a flow of particles carrying positive (or vitreous) electricity. Furthermore, there exist other particles, called positrons, that are identical to electrons in almost every respect except that they carry positive electric charge. Thus, in a deep sense Du Fay was correct in taking a symmetrical view of the two types of electric charge: Positive and negative (or resinous and vitreous) electricity are equally fundamental.

The reader may well also wonder why when amber is rubbed with fur the electrons go from the fur to the amber, but when glass is rubbed with silk the electrons go from the glass to the silk? Oddly enough, we still don't know. The question involves the physics of surfaces of complex solids such as silk or hair, and this branch of physics has still not reached a point where we can make definite predictions with any certainty. In a purely empirical way, there has been developed a list of substances called the triboelectric sequence, part of which goes as follows[10]:

rabbit's fur/lucite/glass/quartz/wool/cat's fur/silk/cotton/wood/amber/resins/metals/teflon.

Substances near the beginning of the list tend to lose electrons, and those near the end of the list tend to collect them. Thus, if two objects are rubbed together, the one closer to the beginning of the list will tend to pick up a positive, or vitreous, electric charge and the one closer to the end will tend to pick up a negative, or resinous, charge. The electrification is most intense for objects that are well separated in the triboelectric sequence. For example, it is easier to electrify amber by rubbing with fur than it is to electrify glass by rubbing with silk. The triboelectric sequence is not well understood theoretically, and even a change in the weather can affect the relative placement of various substances.

It is ironic that we still do not have a detailed understanding of frictional electrification, even though it was the first of all electrical phenomena to be studied scientifically. But that is often the way science progresses—not by solving every problem presented by nature, but by selecting problems that are as free as possible from irrelevant complications and that therefore provide opportunities to get at the fundamental principles that underlie physical phenomena. The study of the electricity produced by friction played a great role in letting us know that there is such a thing as electricity and that it can exert attractive and repulsive forces, but the actual process of electrification by rubbing is just too complicated to provide further insights into the quantitative properties of electricity. By the end of the eighteenth century, the attention of physicists was already beginning to focus on other electrical phenomena.

Electric Discharges and Cathode Rays

The study of electricity widened after Franklin to take in the quantitative details of electrical attraction and repulsion and the connection of electricity with magnetism and chemistry. We will have much to do with these matters later on; but for now, let us follow one line of discoveries, concerning the discharge of electricity through rarefied gases and empty space.

The earliest-known and most spectacular sort of electric discharge is of course lightning. Although the nature of lightning as a current of electricity was demonstrated in 1752 in a celebrated experiment suggested by Franklin, lightning is so sporadic and uncontrollable that its study could reveal little about the nature of electricity. But by the eighteenth century, a more controllable sort of electric discharge was becoming available for scientific study.

In 1709 Hauksbee observed that when the air inside a glass vessel was pumped out until its pressure was about $\frac{1}{60}$ normal air pressure and the vessel was attached to a source of frictional electricity, a strange light would be seen inside the vessel. Flashes of similar light had already been noticed in the partial vacuum above the mercury in barometers. In 1748 Watson described the light in a 32-inch evacuated tube as an "arch of lambent flame." Other observations were recorded by the Abbé Nollet, by Gottfried Heinrich Grummont (1719–1776), and by the great Michael Faraday, about whom more later.

The nature of this light was not understood at first, but today we know that it is a secondary phenomenon. When an electric current flows through a gas, the electrons knock into gas atoms and give up some of their energy, which is then reemitted as light. Today's fluorescent lights and neon signs are

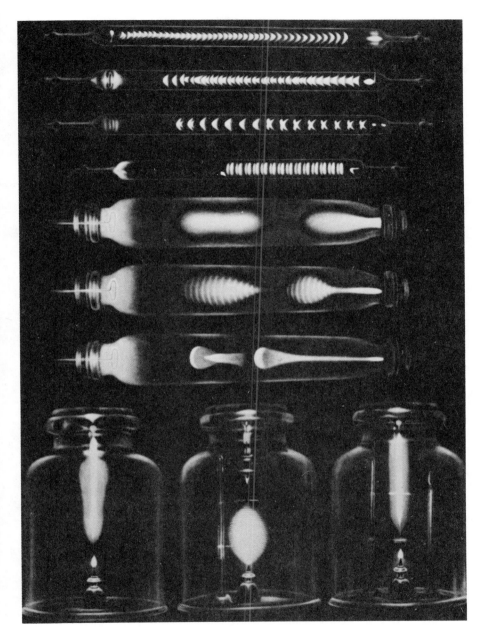

Electrical discharges in gases at low pressure.

based on the same principle, with their color determined by the color of light that is preferentially emitted by the gas atoms: orange for neon, pinkish-white for helium, greenish-blue for mercury, and so on. The importance of the phenomenon for the history of electrical science lay, however, not in the light given off in electric discharges, but in the electric current itself. When electricity collects on an amber rod, or an electric current flows through a copper wire, properties of the electricity are hopelessly mixed up with those of the solid integument of amber or copper. For instance, it would be impossible even today to determine the weight of a given quantity of electricity by weighing an amber rod before and after it is electrified; the weight of the electrons is just too tiny compared with that of the rod. What was needed was to get electricity off by itself, away from the solid or liquid matter that normally carries it. The study of electric discharges in gases was a step in the right direction, but even at $\frac{1}{60}$ atmospheric pressure the air interfered too much with the flow of electrons to allow their nature to be discovered. Real progress became possible only when the gas itself could be removed and scientists could study the flow of pure electricity through nearly empty space.

The turning point came with the invention of really effective air pumps. Early pumps had leaked air through the gaskets around their pistons. In 1885 Johann Heinrich Geissler (1815–1879) invented a pump that used columns of mercury as pistons and consequently needed no gaskets. With Geissler's pump, it became possible to evacuate the air in a glass tube until its pressure was a few ten-thousandths that of normal air at sea level. Geissler's pump was used in 1858–59 in a series of experiments on the conduction of electricity in gases at very low pressure, carried out by Julius Plücker (1801–1868), Professor of Natural Philosophy at the University of Bonn. In Plücker's arrangement, metal plates inside a glass tube were connected by wires to a powerful source of electrcity. (Following Faraday's terminology, the plate attached to the source of positive electricity is called the *anode* and the plate attached to the source of negative electricity is called the *cathode*.) Plücker observed that when almost all air was evacuated from the tube, the light disappeared through most of the tube, but a greenish glow appeared on the glass tube near the cathode. The position of the glow did not seem to depend on where the anode was placed. It appeared that something was coming out of the cathode, traveling through the nearly empty space in the tube, hitting the glass, and then being collected by the anode. A few years later, Eugen Goldstein (1850–1930) introduced a name for this mysterious phenomenon: *Cathodenstrahlen,* or cathode rays.

We know now that these rays are streams of electrons. They are projected from the cathode by electrical repulsion, coast through the nearly empty space within the tube, strike the glass, depositing energy in its atoms which is

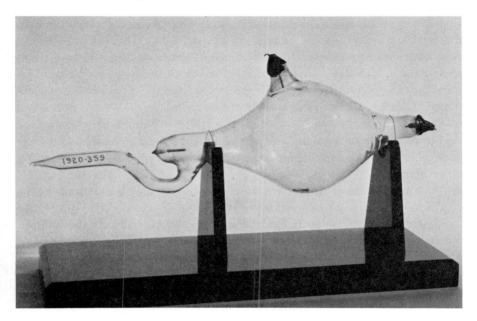

A Crookes tube, made in 1879, devised for studying cathode rays.

then reemitted as visible light, and finally are drawn to the anode, via which they return to the source of electricity. But this was far from obvious to nineteenth-century physicists. Many different clues were discovered, and for a long time they seemed to point in different directions.

Plücker himself was misled by the fact that when the cathode was of platinum, a film of platinum was found deposited on the walls of the glass bulb. He thought that the rays might consist of small pieces of cathode material. We now know that the electrical repulsion felt by the cathode material does indeed cause pieces of the cathode's surface to be torn off (a phenomenon known as sputtering), but this really has nothing to do with cathode rays in general. In fact, Goldstein showed in the 1870s that the properties of cathode rays do not depend on the material of which the cathode is made.

Plücker also observed that the position of the glow on the walls of the tube could be moved by placing a magnet near the tube. As we shall see, this was a sign that the rays consist of electrically charged particles of some sort. Plücker's student J. W. Hittorf (1824–1914) observed that solid bodies placed near a small cathode would cast shadows on the glowing walls of the tube. From this he deduced that the rays travel from the cathode in straight lines. The same phenomena were observed in 1878–79 by the English physicist, chemist, and spiritualist Sir William Crookes (1832–1919), and this led

Crookes to conclude that the rays were molecules of the gas within the tube that had happened to pick up a negative electric charge from the cathode and were then violently repelled by it. (Cromwell Varley, a fellow physicist and spiritualist in Crookes's circle, had already suggested in 1871 that the rays were "attenuated particles of matter, projected from the negative pole by electricity.") But Crookes's theory was effectively refuted by Goldstein, who noted that in a cathode-ray tube evacuated to 1/100,000 normal air pressure, the rays traveled at least 90 centimeters, whereas the typical free path of an ordinary molecule in air at this pressure would be expected to be only about 0.6 centimeters.

A very different theory was developed in Germany on the basis of the observations of the gifted experimentalist Heinrich Hertz (1857–1894). In 1883, while an assistant at the Berlin Physical Laboratory, Hertz showed that the cathode rays were not appreciably deflected by electrified metal plates. This seemed to rule out the possibility that the cathode rays were electrically charged particles, for in that case the ray particles should have been repelled by the plate carrying like charge and attracted to the plate carrying unlike charge. Hertz concluded that the rays were some sort of wave, like light. It was not clear why such a wave should be deflected by a magnet, but the nature of light was then not well understood, and a magnetic deflection did not seem impossible. In 1891 Hertz made a further observation that seemed to support the wave theory of cathode rays: The rays could penetrate thin foils of gold and other metals, much as light penetrates glass.

But the rays were not a form of light. In his doctoral research, the French physicist Jean Baptiste Perrin (1870–1942) showed in 1895 that the rays deposit negative electric charge on a charge collector placed inside the cathode-ray tube. We now know that the reason Hertz had not observed any attraction or repulsion of the rays by electrified plates is that the ray particles were traveling so fast, and the electric forces were so weak, that the deflection was just too small to observe. (As Hertz recognized, the electric charge on his plates was partly canceled by effects of the residual gas molecules in the tube. These molecules were broken up by the cathode rays into charged particles, which were then attracted to the plate of opposite charge.) But as Goldstein has shown, if the rays are charged particles, these particles cannot be ordinary molecules. So what were they?

It is at this point that J. J. Thomson enters the story. Thomson first attempted to measure the speed of the rays. In 1894 he obtained a value of 200 kilometers per second (1/1,500 the speed of light), but his method was faulty and he later abandoned this result. Then in 1897 Thomson succeeded where Hertz had failed: He detected a deflection of the cathode rays by electric forces between the rays and electrified metal plates. His success in this was due largely to the use of better vacuum pumps, which lowered the pressure inside the cathode-ray tube to the point where effects of the residual gas within the tube became negligible. (Some evidence for electrical deflection was found at about the same time by Goldstein.) The deflection was toward the positively charged plate and away from the negatively charged one, confirming Perrin's conclusion that the rays carry negative electric charge.

The problem now was to learn something quantitative about the nature of the mysterious negatively charged particles of the cathode rays. Thomson's method was direct: He exerted electric and magnetic forces on the rays and measured the amount by which the rays were deflected.* To understand how Thomson analyzed these measurements, we must first consider how bodies move under the influence of forces in general.

Flashback: Newton's Laws of Motion

The laws of motion of classical physics were set out by Sir Isaac Newton at the beginning of his great work, the *Principia*.[11] Of these, the key principle is contained in the Second Law, which can be paraphrased as the statement that the force required to give an object of definite mass a certain acceleration is proportional to the product of the mass and the acceleration. To understand what this law means, we have to understand what are meant by acceleration, mass, and force.

Acceleration is the rate of change of velocity. That is, just as the velocity is the ratio of the distance traveled by a moving body to the time that elapses in the motion, acceleration is the ratio of the change in velocity of an accelerating body to the time elapsed during the acceleration. The units in which acceleration is measured are therefore the units of velocity per time, or distance-per-time per time. For instance, falling bodies near the surface of the earth fall with an acceleration of 9.8 meters-per-second per second. This means that after the first second a body dropped from rest in a vacuum will be falling at a

---

* Thomson also used an alternative experimental method, in which he measured the heat energy and electric charge deposited at the end of the tube by the cathode-ray particle and thus avoided the difficult measurement of the deflection of the ray by electric forces. This method was actually more accurate than the one based on the electric and magnetic deflection of the cathode ray. I am describing the electric/magnetic deflection method here first, not because it was historically more important, but because it presents an occasion for a review of electric forces, which I will need to pin down the definition of electric charge. Thomson's other method I will describe below, after a review of the concepts of energy and heat.

speed of 9.8 meters per second, after two seconds at a speed of 19.6 meters per second, and so on.*

The *mass* of a body is the quantity of matter it contains, irrespective of its shape, size, or composition. This is a terribly imprecise definition, but some imprecision is unavoidable here because in classical physics there is nothing more fundamental in terms of which mass could be defined. The definition can be made a little more precise by specifying that when bodies are brought together, as long as they do not produce changes in each other, the mass of the set of bodies is the sum of the masses of the individual bodies. Hence we can often calculate the mass of a complicated system from the masses of its constitutents by just adding them up. The unit of mass most commonly used in science is the gram (g), defined originally as the mass of a cubic centimeter of pure water at normal atmospheric pressure and a temperature of 4° Celsius. One kilogram (kg) is 1,000 grams; one milligram (mg) is 0.001 grams. Since 1875 the kilogram has been defined as the mass of a bar of platinum-iridium alloy held at the International Bureau of Weights and Measures at the Pavillion de Breteuil near Paris, and the gram is defined as a thousandth of a kilogram.

Force is the amount of pushing or pulling, irrespective of its duration or the nature of the body on which the force acts. This, too, is an extremely imprecise definition. It can be made more precise by specifying that two forces are equal if a body acted on in opposite directions by the two forces remains at rest, and that when a number of equal forces act in the same direction on a body the total force is the number of forces times the magnitude of the individual forces. The unit of force could be described as the force exerted by some standard spring extended some standard distance. In that case, since there would be no relation between the units used for force and those used for mass and acceleration, Newton's Second Law would have to be expressed, as it was above, as the statement that the force required to give a body a certain acceleration is simply proportional to the product of the mass of the body and its acceleration.

However, it is possible and much more convenient to relate the definition of the unit of force to the units of mass and acceleration. For instance, if we measure accelerations in meters per second per second and masses in kilo-

---

*Since the numerical value of the velocity is the ratio of the numerical values of a distance and a time, and the numerical value of the acceleration is the ratio of the numerical values of a velocity and a time, it is convenient to replace the word *per* with a fraction sign and say that the units of velocity are length/time—for instance, centimeters/second or miles/hour—and that the units of acceleration are (distance/time)/time, or equivalently distance/time^2—for instance, centimeters/second^2 or miles/hour^2. Thus, the acceleration of falling bodies near the surface of the Earth would be written as 9.8 meters/second^2, or 9.8 m/sec^2 for short.

grams, then we should take as the unit of force the *newton* (N), defined as the quantity of force that would give a mass of 1 kilogram an acceleration of 1 meter per second per second. In this system of units, Newton's Second Law takes the simple form

$$\begin{matrix} \text{Force on a body} \\ \text{required to give it a} \\ \text{certain acceleration} \end{matrix} = \begin{matrix} \text{Mass of} \\ \text{the body} \end{matrix} \times \begin{matrix} \text{Acceleration} \\ \text{of the body} \end{matrix}$$

For a mass of 1 kilogram and an acceleration of 1 meter per second per second, this is just the definition of the newton. The formula holds for all other values of the mass and acceleration because Newton's Law tells us that the force is proportional to both. For instance, if the mass is 2 kilograms and the acceleration is 3 meters per second per second, then the force must be 2×3 times greater than it is for a mass of 1 kg and an acceleration of 1 m/sec^2, so it is 6 newtons. (See Appendix A.)

Some additional comments on Newton's Second Law:

• When other units are used for mass or acceleration, we can still continue to use Newton's Second Law in the above simple form, but then we must use other units of force. For instance, Newton's Law tells us that the force required to give a mass of 1 gram ($= 10^{-3}$ kg) an acceleration of 1 centimeter per second per second ($= 10^{-2}$ m/sec^2) is

$$(10^{-3} \text{ kg}) \times (10^{-2} \text{ m/sec}^2) = 10^{-5} \text{ N.}$$

This unit of force is known as the dyne. Force is still equal to mass times acceleration if we express accelerations in centimeters per second per second, masses in grams, and forces in dynes.

• It is important to distinguish between mass and weight. Weight is a kind of force, the force that is exerted on a body by gravity. It has already been mentioned that bodies near the earth's surface fall with an acceleration of 9.8 meters per second per second; hence Newton's Law shows that a mass of 1 kilogram has a weight of 9.8 newtons. Similarly, Newton's Law says that a mass of m kilograms has a weight of 9.8m newtons. The fact that all bodies fall with the same acceleration thus implies that weight is proportional to mass. (This fundamental property of gravitation provided the clue that led Einstein to the General Theory of Relativity.) When we put an object on a scale, we are really measuring its weight and not its mass; a reading of m kilograms really means that the weight is 9.8m newtons. This distinction becomes important if we imagine weighing objects elsewhere than on the earth's surface. For instance, a mass of 1 kilogram weighs 9.8 newtons on the surface of the earth. On the surface of the moon its mass would still be 1 kilogram, but in the weaker lunar gravity its weight would be only 1.62 newtons.

• Although Newton's Second Law is used to define our units of force, the law itself is not merely a definition of force. Even without a precise independent definition, we have an intuitive notion of force that gives content to the Second Law. For instance, it is not a mere definition to say that if a certain stretched spring gives a certain mass a certain acceleration it will give twice the mass half the acceleration, and that two such springs acting in the same direction will give the mass twice the acceleration. Also, a constant force acting on a body will give it a constant acceleration, so the body's velocity will increase by the same amount in each second. Experimental facts of this sort provide the basis for the Second Law.

• As a special case of Newton's Second Law, a body of nonvanishing mass, when acted on by zero force, will experience zero acceleration—that is, it will move with constant velocity. Newton listed this separately as the First Law of Motion. The Third Law of Motion is that action equals reaction: If one body exerts a force on another, then the second body exerts an equal force in the opposite direction on the first.

• It is correct to define velocity as the ratio of distance traveled to elapsed time only when the velocity is constant, and it is correct to define acceleration as the ratio of change in velocity to elapsed time only when the acceleration is constant. Otherwise, these ratios give the average velocity and the average acceleration, respectively. When the velocity or acceleration are changing, we can define the instantaneous velocity or acceleration at any moment as the average values of velocity or acceleration over a vanishingly small time interval centered on that moment. Newton's Law actually relates the force to the instantaneous acceleration.

• Velocity, acceleration, and force are *vectors*—that is, they have direction as well as magnitude. It is often convenient to describe such quantities in terms of their components along specified directions. For instance, when we specify the velocity of a ship by saying that it has an eastward component of 10 kilometers/hour and a northward component of 15 kilometers/hour, we mean that in one hour it moves 10 km toward the east and 15 km toward the north. (Such a ship is actually moving at about 18 km/hr in a direction approximately northeast by north.) Likewise, when we specify that the acceleration of a ship has an eastward component of 2 km/hr^2 and a northward component of 1 km/hr^2 we mean that, whatever velocity it had originally, in each hour its

THE DISCOVERY OF THE ELECTRON

eastward component increases by 2 km/hr and its northward component increases by 1 km/hr. Forces can similarly be described in terms of components, which give the amount of pushing or pulling along specified directions.

Components of vectors can be negative as well as positive; for instance, if the eastward component of the velocity is −20 km/hr then in each hour the ship moves 20 km to the *west*, and if the eastward component of the acceleration is −2 km/hr per hour then in each hour the eastward component of velocity decreases (or the *westward* component *increases*) by 2 km/hr. A force with a negative eastward component is actually pushing toward the west. (In these examples, the motion is purely horizontal, so only two components are needed to specify a velocity or an acceleration or a force. In general, three components are needed—for instance, east, north, and up.) Newton's Second Law applies separately to each component of force and acceleration—it says that *the component of force in any direction is equal to the mass times the corresponding component of acceleration.*

• When several forces act on an object, the total force is the sum of the individual forces. To be more specific, each component of the total force is the sum of the corresponding components of the individual forces. For instance, if a body is acted on by one force that has a northward component of 3 newtons and an eastward component of 1 newton, and by a second force that has a northward component of −1 newton and an eastward component of 6 newtons, the total force has a northward component of 2 newtons and an eastward component of 7 newtons.

Deflection of Cathode Rays

Thomson used Newton's Second Law to obtain a general formula that would allow him to interpret measurements of the cathode-ray deflection, produced in his experiment by various electric or magnetic forces, in terms of the properties of the cathode-ray particles. In his cathode-ray tube, the ray particles pass through a region (let's call it the deflection region) in which they are subjected to electric or magnetic forces acting essentially at right angles to their original direction, and then through a much longer force-free region (the drift region) in which they coast freely until they hit the end of the tube. A glowing spot of light appears where the ray particles hit the glass wall at the end of the tube, so it was easy for Thomson to measure the displacement of the ray produced by the forces acting on it by measuring the distance between the locations of the glowing spot when the forces were on and when the forces were turned off.

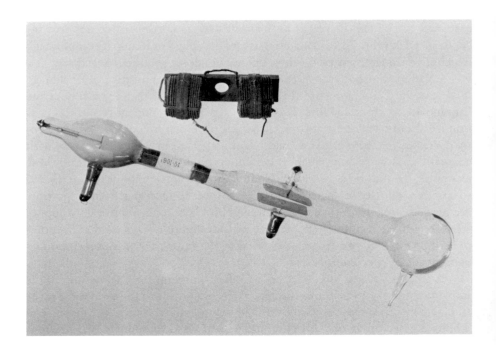

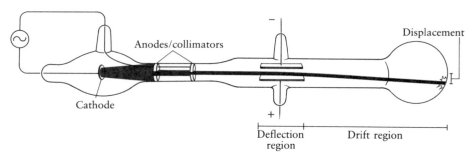

Above: One of the tubes with which J. J. Thomson measured the mass-to-charge ratio of the electron. Below: A schematic view of Thomson's apparatus. The cathode is connected by a wire through the glass tube to a generator that supplies it with negative electric charge; the anode and collimator are connected to the generator by another wire so that negative electric charge can flow back to the generator. The deflection plates are connected to the terminals of a powerful electric battery, and are thereby given strong negative and positive charges. The invisible cathode rays are repelled by the cathode; some of them pass through the slits in the anode and collimator, which only admit a narrow beam of rays. The rays are then deflected by electric forces as they pass between the plates; they then travel freely until they finally hit the glass wall of the tube, producing a spot of light. (This figure is based on a drawing of Thomson's cathode-ray tube in Figure 2 of his article "Cathode Rays," *Phil. Mag.* 44(1897), 293. For clarity, the magnets used to deflect the rays by magnetic forces are not shown here.)

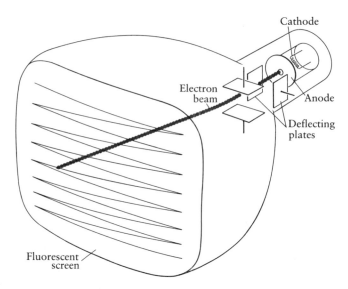

A schematic view of a more familiar cathode-ray tube, the modern television picture tube. As we have seen, Thomson used the position of the glowing spot where the cathode ray hit the end of the tube to tell him about the path taken by the ray, which was invisible as it passed through the vacuum of his tube. Since Thomson's time, this glowing spot has become much more familiar to all of us as the basis of television. A television picture tube is essentially just a cathode-ray tube aimed at the viewer. In it, the cathode ray is steered by electric forces so that it passes regularly back and forth over the end of the tube. When the ray hits the screen of specially coated glass at the end of the tube, a spot of light appears. The television signal controls the strength of the cathode ray as it strikes each spot on the screen, so that a pattern of light and dark appears successively on the screen. The eye and brain respond slowly, and see this pattern as an instantaneous picture.

(See the schematic diagram of Thomson's apparatus.) Thomson's formula states that

$$\text{Displacement of ray at end of tube} = \frac{\text{Force on ray particle} \times \text{Length of deflection region} \times \text{Length of drift region}}{\text{Mass of ray particle} \times \left(\text{Velocity of ray particle}\right)^2}$$

To take an illustration using numbers that are more or less realistic, suppose that the force exerted on the ray particles is 10^{-16} newtons, the length of the deflection region is 0.05 meters, the length of the drift region is 1.1 meters, the mass of the cathode-ray particles is 9×10^{-31} kilograms, and the velocity of the cathode-ray particles is 3×10^7 meters per second. Then the displacement of the ray when it hits the end of the tube is

$$\text{Displacement} = \frac{(10^{-16}\,\text{N}) \times (0.05\,\text{m}) \times (1.1\,\text{m})}{(9 \times 10^{-31}\,\text{kg}) \times (3 \times 10^7\,\text{m/sec})^2} = 0.0068\,\text{m}\,.$$

This is about a quarter inch, not a difficult distance to measure. (The answer comes out in meters because we are using a coherent system of units, in which all lengths are expressed in meters, all times in seconds, all masses in kilograms, all velocities in meters per second, all forces in newtons, and so on. We could have used any coherent system of units—since a displacement is a length, the answer would always have come out in whatever units of length were used in that system.)

Thomson's formula is derived algebraically in Appendix B. However, even without algebra, it is easy to see why it takes the form it does. The important thing to keep in mind is that the forces exerted on the cathode-ray particles give them an acceleration at right angles to the axis of the tube, so that by the time the particles emerge from the deflection region they have a small component of velocity at right angles to their original motion. This component is equal to the product of the acceleration and the time they spend in the deflection region. For definiteness, suppose that the tube is horizontal and the deflection is downward, as in the diagram. Next the ray particles enter the drift region, and since there are no forces acting on them they keep the same horizontal and downward components of velocity. Since the distance traveled in any direction equals the component of velocity in that direction times the elapsed time, the downward displacement of the ray when it hits the end of the tube is simply equal to the downward component of velocity produced in the deflection region times the length of time it spends in the drift region. (We are ignoring the displacement of the ray while it is in the deflection region, because this region is much shorter than the drift region; the particles spend much less time in it, so the displacement that takes place there is relatively quite small.) Putting this together, we see that the displacement of the ray when it hits the end of the tube is equal to its downward acceleration in the deflection region times the time the ray particles spend in the deflecting region (this product gives the downward velocity) times the time they spend in the drift region. The time they spend in each region is just the length of the region divided by the (unchanged) horizontal velocity of the ray particles; this is why the lengths of the deflection and drift regions appear in the numerator of Thomson's formula and why the ray velocity appears twice (that is, squared) in the denominator. Finally, Newton's Second Law of Motion says that the component of acceleration in any direction equals the force in that direction divided by the mass; this is why the force appears in the numerator and the mass in the denominator of Thomson's formula.

In his experiment, Thomson measured the displacement produced by various electric or magnetic forces acting on the ray. What can this reveal about the cathode-ray particles? Of the quantities that appear in Thomson's formula, the lengths of the deflection and drift regions are known quantities determined by the design of the cathode-ray tube; the ray particles' mass and velocity are properties of the ray particles that one would like to determine. But what about the force? As we will soon see, the electric force acting on a particle is proportional to the electric charge carried by the particle. Referring back to Thomson's formula, we see that the displacement of the ray when it strikes the end of the tube is proportional to a particular combination of parameters of the ray particles—their electric charge divided by their mass and by the square of their velocity—and therefore measurements of the ray displacement can yield a value for only this combination of parameters. But this is not really what one wants to know. The interesting quantities are the charge and the mass of the ray particles; the velocity is just whatever velocity happens to be produced in a particular cathode-ray tube.

Thomson was able to get around this difficulty by also measuring the deflection caused by a magnetic force. We will soon see that, unlike the electric force, the magnetic force acting on a particle is proportional to the particle's velocity as well as to its electric charge. Therefore, the displacement of the ray by magnetic forces depends on a different combination of ray-particle parameters than the displacement due to electric forces. By measuring the deflections due to electric and to magnetic forces, Thomson was able to learn the values of two different combinations of ray-particle parameters, and in this way he could determine both the ray-particle velocities and the ratio of their charge and mass.

Thomson's results will be described later in this chapter, but before we come to them we must say something about the theory of electric and magnetic forces and we must calculate the deflection that they produce in the cathode ray.

Flashback: Electric Forces

In order to use measurements of the electric deflection of cathode rays to learn about the properties of cathode-ray particles, Thomson had to be able to calculate the electric force on these particles. We will now take a look at the quantitative theory that describes these forces, and at how it developed.

Early speculation about electric forces relied heavily on an analogy with Newton's theory of gravitational forces. At the end of the *Principia*, Newton described gravitation as a cause that acts on the sun and the planets "according to the quantity of solid matter which they contain and propagates on all sides to immense distances, decreasing always as the inverse square of the distances." That is,

$$\frac{\text{Gravitational force exerted by particle 1 on particle 2}}{} = \frac{G \times \frac{\text{Mass of particle 1}}{} \times \frac{\text{Mass of particle 2}}{}}{\left(\text{Distance between particles 1 and 2} \right)^2}.$$

where G is a fundamental constant whose value depends on the system of units used to describe forces, masses, and distances; this value must be found from experiment. (Modern measurements tell us that, with forces in newtons, masses in kilograms, and distances in meters, $G = 6.672 \times 10^{-11}$.) Most of the details of Newton's law have an intuitively satisfying plausibility. The force by which one body attracts another is naturally proportional to both their masses, so for instance if one or the other mass is doubled then so is the force, and it naturally decreases as the bodies are moved further apart. It was irresistible to guess that the electric force might obey a similar law, also proportional to the inverse square of the distance, but with electric charge playing the role that mass plays in gravitational forces.

The first attempt to measure the distance dependence of the electric force was made in 1760 by the Swiss physicist Daniel Bernoulli (1700–1782). Bernoulli's apparatus was primitive, and it is not clear whether he really discovered an inverse-square law of electrical attraction and repulsion or merely checked that his observations were consistent with this preconceived law.

The inverse-square law was guessed at on rather indirect grounds by the English physicist and chemist Joseph Priestley (1733–1804), the discoverer of oxygen. Priestley observed that a body placed inside a closed electrified metal cavity would feel no electric force, even when it was placed close to one of the walls of the cavity. This was suggestive of a result found by Newton: that, as a consequence of the proportionality of gravitational forces to the inverse square of the distance, a body inside a hollow, massive spherical shell would feel no gravitational attraction to the walls of the shell. But the analogy was not a very good one. For gravity the absence of force inside a shell depends critically on the spherical symmetry of the shell, whereas the absence of electric forces inside a metal cavity arises in part because of the way that electric charges distribute themselves on the metal surface, and applies whatever the shape of the cavity.

Direct experimental tests of the inverse-square law for electric force were carried out in 1769 by John Robison (1739–1805) for repulsion only, and in 1775 in unpublished work by Henry Cavendish, after whose family Thomson's laboratory in Cambridge was named. However, the first really persuasive experimental tests were made in 1785 by Charles Augustine Coulomb (1736–1806).

Coulomb was a military engineer who learned his trade and wrecked his health while supervising the construction of Fort Bourbon in Martinique in 1764–1772. Back in France, Coulomb was able to carry out exhaustive experiments on friction at the shipyards at Rochefort, and in 1781 was elected to the Académie des Sciences. This affiliation gave him the opportunity to settle in Paris and to devote most of his time to research. He published the results of his researches on electricity and magnetism between 1785 and 1791 in seven memoires to the Academy.

Coulomb used a sensitive device of his own invention, the torsion balance, to measure the forces between small pith balls. The inverse-square law was found to hold accurately for various charges and separations; for instance, reducing the separation between balls by half increased the force between them by a factor of 4. Coulomb also stated that the force between electrically charged bodies is proportional to the product of the electric charges (in Coulomb's term, the "electrical masses"), as might be expected by analogy with the gravitational law of force. That is,

$$\begin{array}{l}\text{Electric} \\ \text{force exerted} \\ \text{by particle 1} \\ \text{on particle 2}\end{array} = \dfrac{k_e \times \begin{array}{c}\text{Electric charge} \\ \text{of particle 1}\end{array} \times \begin{array}{c}\text{Electric charge} \\ \text{of particle 2}\end{array}}{\left(\text{Distance between particles 1 and 2}\right)^2},$$

where k_e is, like G, a fundamental constant that depends on the units used to define forces, charges, and distances and must be determined by experiment.

In one check of the dependence of the electric force on the product of charges, Coulomb measured the force between two electrically charged pith balls at a definite separation, then removed one of the balls and touched it to another uncharged ball of equal size, so that its charge was equally shared with the other ball and therefore reduced by half. When this ball was placed back in its original position, the force between it and the other electrically charged ball was found to be reduced by half, as would be expected from Coulomb's law.

Force is a directed quantity, or vector, so it is also necessary to say something about the direction of the electric force. I do not know if Coulomb

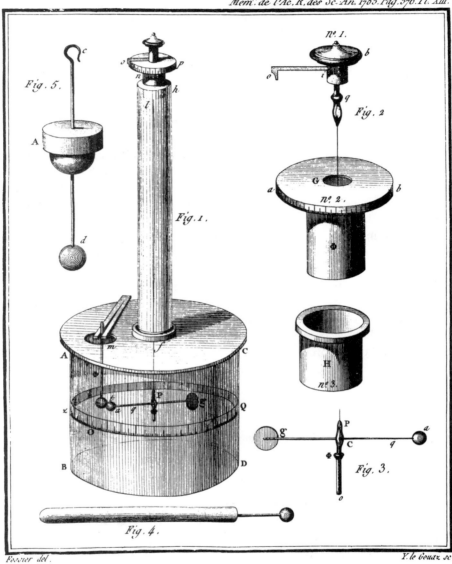

Fig. 5.

Fig. 1.

nº 1.

Fig. 2.

nº 2.

nº 3.

Fig. 3.

Fig. 4.

A 1785 etching of Coulomb's torsion balance, with which he proved the inverse-square law for electrical attraction.

ever stated it explicitly, but it is almost obvious that the electric force acts along the line separating the two charges. (There is no other special direction along which one could imagine it would act.) If we adopt the convention that a repulsive force is positive and an attractive force is negative, then Du Fay's observation that like charges repel and unlike charges attract can be summarized in the simple statement that k_e is a positive number.

What units should we use for electric charge? There is a "practical" system of electric units in which the fundamental unit is a quantity of electric current, the *ampere*. The original definition of the ampere was based on the magnetic forces between electric currents, but for the present we can think of the ampere as the current that will just blow out a one-ampere fuse. The practical unit of electric charge is the *coulomb* (C), defined as the electric charge that passes a given point in one second in a wire that carries a one-ampere current. (That is, an ampere is a coulomb per second.) With forces in newtons, distances in meters, and charges in coulombs, k_e has the measured value 8.987×10^9 N m^2/C^2. (It is also possible to adopt a unit of electric charge known as the *electrostatic unit,* or *statcoulomb,* which is defined so that the constant k_e has the value 1. However, this is not the most frequently used unit of electric charge, and we will use only the practical system.)

It is very convenient to restate Coulomb's law in modern terms, first used in this way by James Clerk Maxwell. The electric force on any body is always proportional to the electric charge of the body. We can call the factor of proportionality the *electric field,* so that

$$\text{Electric force on a body} = \begin{matrix}\text{Electric charge} \\ \text{of the body on} \\ \text{which the force acts}\end{matrix} \times \begin{matrix}\text{Electric} \\ \text{field}\end{matrix}$$

The electric field introduced in this way clearly depends on where the body is placed and on all the electric charges and distances of all the other bodies that produce the electric field, but it does not depend on the nature of the body on which the force acts or on its charge. For instance, when the electric force on one body is exerted by one other body, Coulomb's law can be reinterpreted as follows:

$$\begin{matrix}\text{Electric field} \\ \text{due to} \\ \text{a charged body}\end{matrix} = \frac{k_e \times \begin{matrix}\text{Electric charge of} \\ \text{body producing field}\end{matrix}}{\left(\begin{matrix}\text{Distance from body} \\ \text{producing field}\end{matrix}\right)^2}.$$

That is, combining these two rules simply gives back Coulomb's law in its original form.

The units of electric field are those of force per charge, that is, newtons per coulomb.* Like force, electric field is a directed quantity; the electric force on a charged body is in the same direction as the field if the charge is positive; in the opposite direction if the charge is negative. Also, the electric field produced by a charged body points away from the body if the charge is positive; toward the body if the charge is negative. The electric field produced by a set of charges is the vector sum of the electric fields that would be produced by the individual charges; that is, each component (north, east, up) of the total electric field is the sum of the corresponding components of the individual fields.

The introduction of the concept of electric field marked a shift away from Newton's idea of a force as an influence one body exerts directly on another body some distance away. Instead, one thinks of the electric field at a given position as a condition of space at that position, which acts directly on any charged body at that position and receives contributions from all electric charges at all other positions. Increasingly in modern physics, fields have taken on the character not of mere mathematical artifices that help us to calculate the forces between particles, but as physical entities in their own right—inhabitants of our universe that may in fact be more fundamental than the particles themselves.

A pictorial representation of electric fields devised originally by Michael Faraday (1791–1867) gives a good intuitive sense of how these fields behave, and even can be used to calculate the fields in simple cases, like that of Thomson's cathode-ray tube (see Appendix C). Draw lines throughout space, the lines at any point pointing in the direction of the electric field at that point. Make the number of lines that pass through a small area perpendicular to the electric field at a given point to be equal to the area times the magnitude of the field at that point.† For a single point charge, the lines will then be everywhere directed away from the charge (or toward it if the charge is negative), and the number of lines through a sphere around the charge will equal the sphere's area times the field. But the area of the sphere is proportional to the square of the radius, and we have just seen that the electric field on the

---

* This unit is more commonly known as a volt per meter, for reasons that will be explained later in this chapter.

† With this definition, the number of lines of force depends on the units used to describe the electric field—for instance, it is very different if we give the electric field in dynes per statcoulomb, or newtons per coulomb, or something else. This serves to emphasize that the field lines are not real, and that no absolute significance can be given to the numbers of field lines, only to their direction and to the *relative* numbers at different points.

Michael Faraday.

surface of a sphere with the charged body that produces the field at the center will be inversely proportional to the square of the radius. Thus, when we calculate the number of lines passing through the sphere, the radius of the sphere cancels out: The number is independent of the radius of the sphere. Since the number of lines passing through one sphere around a charged particle is the same as for any other such sphere, lines do not begin or end anywhere in charge-free space. Furthermore, the field for an arbitrary configuration of electric charges is the sum of the fields produced by the individual charges, and in consequence this property of the field lines will be true in general.

The point of all this is not to take a known field configuration and reexpress it in terms of field lines, but rather to learn how to calculate the electric field in various situations from intuitively plausible properties of the field lines. We will see how this works in the next section.

Electric Deflection of Cathode Rays

In Thomson's experiment, electric forces were produced by parallel, charged metal plates (see the diagram on p. 30). As we have seen, the electric force on any charged body can in general be expressed as the product of the charge and the value of the electric field at the position of the body. Thus, in order to interpret measurements of cathode-ray deflection in terms of the properties of the cathode-ray particles, it was necessary for Thomson to be able to determine the electric field along the path of the ray between the plates.

This problem is simplified tremendously if one takes account of the fact that in Thomson's experiment the length and width of the metal plates were much greater than their separation. In consequence, at most points between the plates one can ignore any effects of the plate edges. Thus, except near the edges of the plates, the electric field between the plates must lie at right angles to the plates (pointing from the positive plate to the negative plate), because there is no other special direction in which it could be expected to point, as the diagram on p. 42 shows. Also, the electric field cannot depend on location along the plates, because any point on the plate is like any other. (Even if we put a nonuniform distribution of electric charge on the plates, the electric forces that would be set up by these charges would move the charges around in the metal plates until their distribution was uniform.) Finally, and perhaps most surprising, the electric field at a point between the plates also cannot depend on the distance from this point to either plate. This is a consequence of the interpretation of the electric field as the number of field lines per unit area. A glance at the diagram on p. 42 shows that the same number of field lines pass

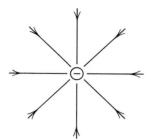

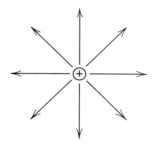

Electric field lines
for an isolated negative charge.

Electric field lines
for an isolated positive charge.

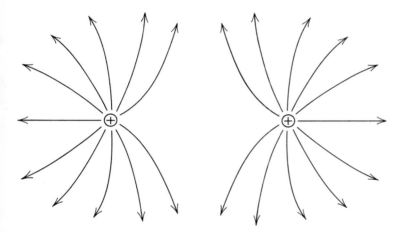

Electric field lines for a pair of positive charges.

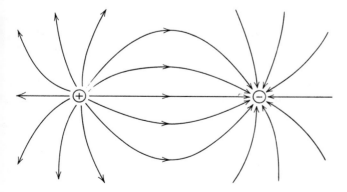

Electric field lines for a pair of unlike charges.

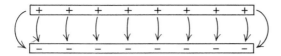

Electric field lines between a pair of parallel, oppositely charged metal plates.

through a given area placed at right angles to the field lines anywhere between the plates, irrespective of how far the area is from either plate. We conclude then that the electric force in Thomson's problem is indeed at right angles to the axis of the cathode-ray tube, and has a magnitude equal to the charge of the electron times a constant, the electric field. Using the results quoted on p. 31 we see that electric forces produce a displacement of the cathode ray at the end of the tube given by the following formula:

$$\text{\shortstack{Displacement\\of cathode\\ray by\\electrical field}} = \frac{\text{\shortstack{Charge of\\ray particle}} \times \text{\shortstack{Electric\\field}} \times \text{\shortstack{Length of\\deflection region}} \times \text{\shortstack{Length of\\drift region}}}{\text{\shortstack{Mass of\\ray particle}} \times \left(\text{\shortstack{Velocity of\\ray particle}}\right)^2}.$$

In order to use measurements of this deflection to learn something about the ray particles, it is necessary to know the value of the electric field between the charged metal plates. One way to find it is to place a test particle of known electric charge between the plates and measure the force on it; the electric field is then the ratio of this force to the charge of the test body. The electric field can also be determined from the known voltage of the battery used to charge the metal plates and from their separation. This is the method actually used by Thomson, but we will have to come back to this later, after we have reviewed what is meant by voltage; for now, we will simply regard the electric field as a quantity that has been determined by some means or other.

As we have already seen, a measurement of the deflection of cathode rays by electric fields could allow one to determine only the ratio of the charge of the ray particles to the product of their mass and the square of their velocity. To get at the ratio of charge to mass for cathode-ray particles, one needs to know their velocity. Thomson had measured this directly in 1894, but the measurement was faulty, and by 1897 he had decided not to trust it. Instead, he measured the deflection produced by a kind of force that has a different dependence on velocity: the force of magnetism.

Flashback: Magnetic Forces

The phenomenon of magnetism has been known at least as long as that of electricity. Plato's *Timaeus* speaks not only of amber, but also of the "Heraclean stone." This was a lodestone, a naturally magnetized piece of iron ore that can pick up small bits of iron and can also give them the same capability.* Lodestones were known in ancient China, and there are cryptic references to their use as compasses for magical purposes as early as A.D. 83.† A detailed description of a magnetic compass made by floating a small "fish" of magnetized iron in water is found in a Chinese book of 1084.‡ The Chinese were also the first to discover that lodestones have two poles toward which small bits of metal are drawn, and that it is one of these poles (the "north-seeking pole") that is drawn toward the north and the other toward the south.[12]

Knowledge of magnetism lagged in the West, but the polarity of lodestones was noted in 1269 by Pierre de Maricourt (also known as Peter Peregrinus).[13] Maricourt made the fundamental observation that the north-seeking pole of a magnet will repel the north-seeking pole of another magnet, and likewise for two south-seeking poles, but that north-seeking poles attract south-seeking poles.

The foundations of a scientific understanding of magnetism were laid in Elizabethan London by William Gilbert. Relying on Maricourt's observation of the polarity of magnets, Gilbert correctly guessed that this provides the explanation for the magnetic compass. The earth is itself a giant magnet, whose "south-seeking" magnetic pole lies somewhere near the geographic North Pole and attracts the north-seeking pole of whatever magnet is used in a compass. Perhaps most important was Gilbert's recognition that, despite their

---

* The Greek name for this ore was λίθοσ Μαγνῆτισ, or "Magnesian stone," after the city of Magnesia (modern Manisa) in western Asia Minor, where the ore was mined. The modern name is magnetite, or Fe_3O_4. The city of Magnesia gave its name not only to magnets and magnetite, but also to the element magnesium, named after another ore, magnesia, from which it was obtained.

† This is the *Lun Hêng (Discourses Weighted in the Balance)* of Wang Chung. Needham[12] quotes it as referring to a "south-controlling spoon," a piece of lodestone carved in the shape of the constellation we call the Big Dipper or Great Bear; when placed on a polished bronze plate the carved stone will rotate until it points south. It is interesting that magnetic compasses were always described as indicating the southward direction in China, but the northward direction in Europe.

‡ This is the *Wu Ching Tsung Yao (Compendium of Important Military Techniques)* of Tsêng Kung-Liang, quoted by Needham.[12] The iron "fish" in this compass was magnetized not by stroking with a lodestone, but by heating the iron and then holding it fixed in a north–south direction as it cooled.

similarities, magnetism and electricity are different phenomena; a lodestone will attract only iron, but will do so without having to be rubbed, whereas amber will attract small bits of any material, but will do so only after having been electrified by rubbing with some suitable material. But although they are different phenomena, magnetism and electricity are deeply related. We now know that the magnetism of a lodestone or a horseshoe magnet is set up by electric currents within the atoms of iron, and the magnetism of the earth is set up by electric currents flowing in the molten material within the planet. These complicated phenomena are still studied as a means of learning about how iron atoms orient themselves in solids, or how matter moves within the earth, rather than as a means of learning about magnetism itself. After Gilbert, progress toward a fundamental understanding of the nature of magnetism was not made by studying the magnetism of iron or of the earth, but by studying *electromagnetism,* the magnetism produced by controlled macroscopic electric currents.

The credit for discovering electromagnetism belongs to Hans Christian Oersted (1777–1851). The origins of this discovery are not entirely clear. According to one account Oersted, a professor of physics at the University of Copenhagen, noticed during a lecture demonstration early in 1820 that a compass needle was deflected when an electric current was allowed to flow through a nearby wire. Oersted's source of current was a battery, something like an automobile storage battery. (The battery had been invented in 1800 by Count Alessandro Volta (1745–1827), and this invention had given rise to a great deal of experimentation throughout Europe on the properties of electric currents. Oddly, no one before Oersted had noticed the electromagnetic effect.) The electric currents available at first to Oersted were very weak. When he repeated the experiments in July 1820 with a more powerful battery, the results were striking. A compass needle placed near a current-carrying wire would swing until it pointed at right angles to the wire and also to the line between the compass and the wire. If the compass was continuously moved in the direction it pointed, it would trace out a circle around the wire. Reversing the direction of the electric current would reverse the direction of the compass needle. The effects persisted even when plates of glass, metal, or wood were interposed between the wire and the compass. A little later, Oersted showed that the effect was symmetrical: Not only did a current-carrying wire exert forces on a magnet such as a compass needle, but also a magnet would exert a force on a coil of wire carrying an electric current, one end of the coil acting like the north pole of a magnet and the other like the south. So magnetism and electricity were not entirely distinct after all.

It is sometimes supposed that the speed of scientific communication and

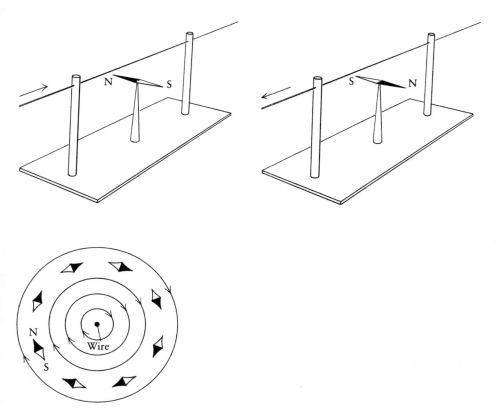

A current-carrying wire exerts a force on a compass needle. The direction of the force depends on the direction of the current. The diagram below shows the direction of the force near a wire in which an electric current flows into the page away from the reader.

the pace of scientific change are much faster now than in previous centuries. But few discoveries have ever had the sudden impact of Oersted's discovery of electromagnetism. His first results were announced on July 21, 1820, in a four-page tract in Latin that was immediately sent to scientific academies throughout Europe.[14] Before the end of the year, translations had appeared in scientific journals in English, French, German, Italian, and Danish.

The most fateful announcements of Oersted's result was made in Paris at the Institut de France on September 11, 1820. In the audience was André Marie Ampère (1775–1836), professor of mathematics at the École Polytechnique. Ampère began a series of experiments, and at the next meeting of the Institut, one week later, he announced a crucial new result: Not only did electric currents exert forces on magnets and magnets exert forces on electric currents, but also electric currents exerted forces on each other. Specifically, parallel wires attracted or repelled each other if they carried electric currents flowing in the same or opposite directions, respectively. Ampère soon reached the conclusion that all magnetism is electromagnetism, and that it is the tiny electric currents circulating in the particles of a lodestone that give it its magnetic properties.

The detailed properties of electromagnetism were worked out on the basis of further experiments and mathematical analysis by Ampère and (using a somewhat different approach) by Jean-Baptiste Biot (1774–1862) and Félix Savart (1791–1841). The simplest case is that of two long parallel wires carrying electric currents. As Ampère found, the force that one wire exerts on the other has a magnitude given by the formula

$$\frac{\text{Force exerted}}{\substack{\text{by wire 2} \\ \text{on wire 1}}} = \frac{2k_\text{m} \times \frac{\text{Current}}{\text{in wire 1}} \times \frac{\text{Current}}{\text{in wire 2}} \times \frac{\text{Length of}}{\text{wires}}}{\text{Distance between wires}}.$$

Here k_m is another universal constant, which depends on the units used to measure forces and electric currents.* The ampere is defined in such a way that

---

* Since k_m is defined by this formula, why did we not define a different constant, say K_m, to have the value $2k_\text{m}$, and write this formula in terms of K_m, without any extraneous factor of 2? The reason is that if we were to eliminate the factor 2 here in this way, it would reappear in a great many other places. For instance, the force between two *short* parallel segments of current-carrying wire separated by a distance much larger than their length would then include an extra factor of $\frac{1}{2}$.

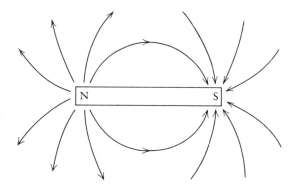

Magnetic field lines around a bar magnet.

$k_m = 10^{-7}$ if electric currents are measured in amperes and forces are measured in newtons.*

To deal with more complicated situations, instead of writing a generalized formula for forces between current elements it is convenient to follow an approach like that used for electricity and introduce the concept of the magnetic field. The direction of the magnetic field at any point is simply defined as the direction of the magnetic force that would be felt by the north-seeking pole of a magnet at that point. Around a lodestone or other permanent magnet, the magnetic field points away from the north-seeking pole (because like poles repel) and toward the south-seeking pole (because unlike poles attract). Also, as Oersted discovered, at a point near a long straight current-carrying wire, the magnetic field points at right angles to the wire and to the line between this

---

* This is the original definition of the ampere; for practical purposes it has been supplanted in part by a definition in terms of electrolysis, which will be discussed in Chapter 3. There is another unit, the abampere or electromagnetic unit (emu), which is defined so that the force per unit length between two long wires carrying currents of 1 abampere at a distance of 1 cm is two dynes per cm. (That is, $k_m = 1$ if forces are measured in dynes and currents are measured in abamperes.) From this, it is easy to work out that 1 abampere = 10 amperes. The ampere and related units like the coulomb and the volt were introduced some time ago as part of a move (in my opinion, misguided) to replace the electromagnetic units based on the centimeter, the gram, and the second with supposedly more practical units.

point and the wire.* (See the diagram on p. 45.) Now, as already mentioned,
Ampère had found that the magnetic force on a second parallel wire is along
the line from one wire to the other and perpendicular to them both—in other
words, the force is perpendicular to the wires and to the magnetic field. This is
the general rule—the magnetic force on any segment of current-carrying wire
always acts in a direction perpendicular to both the magnetic field and the
wire. (See the diagram on p. 49)

The force on a segment of wire in a given magnetic field is proportional
to the current carried by the wire and to the length of the segment. It also
depends on the angle between the field and the wire, vanishing when they are
parallel and reaching its maximum when they are at right angles. Therefore,
we can define the magnitude of the magnetic field by specifying that the force
exerted on a wire segment held at right angles to the magnetic field is given by
the following formula:

$$\text{Force on wire} = \frac{\text{Current carried}}{\text{by wire}} \times \frac{\text{Length}}{\text{of wire}} \times \frac{\text{Magnetic}}{\text{field}}$$

The units of magnetic field are, then, those of force per current per length, for
instance, newtons/ampere meter.† The magnetic field of the earth is about
5×10^{-5} N/amp m, the magnetic field in interstellar space is typically about
10^{-9} N/amp m, and the strongest magnetic field that can be steadily main-
tained in modern laboratories is about 10 N/amp m.

We can put together what we have learned so far to deduce a formula
for the magnetic field produced by a straight length of wire carrying an electric
current. Return for a moment to the case of two parallel wires, in which the
magnetic field produced by one wire is in a direction perpendicular to the other
wire. If we demand that the force exerted on one wire by the current in a
second wire (as given by the formula on p. 46) be nothing but the force
exerted by the magnetic field produced by the second wire (as given by the

---

* Ampère gave a convenient rule for determining the direction of the magnetic field produced by a
long straight current-carrying wire: Imagine a little man who swims along the wire in the direction
of the electric current and faces the point where the field is to be measured. Then the field points in
the direction of the swimmer's left arm. Another way of expressing this rule: If you place the
thumb of your right hand on the wire, pointing in the direction of the current, then your fingers
will curl in the direction of the magnetic field.

† This unit is also known as a weber/m² for reasons that need not concern us here. Another unit of
magnetic field, the gauss, is in common use among physicists. It is defined so that 1 gauss equals
10^{-4} newton per ampere meter. The gauss is so widely used as the unit of magnetic field that the
process by which the slight magnetization picked up by submarines from the earth's magnetic field
is removed between tours of duty is known in the U.S. Navy as "degaussing."

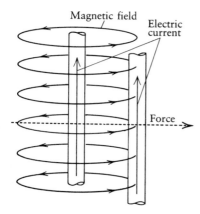

Magnetic field
Electric current
Force

Magnetic force exerted by one electric current
on another, parallel one.

formula on p. 48), we can see that the magnetic field produced by the second
wire is given by the formula

$$\frac{\text{Magnetic field due}}{\text{to current in wire}} = \frac{2k_m \times \text{Current}}{\text{Distance from wire}}.$$

For instance, at a distance of 0.02 meters (about one inch) from a long wire
carrying a current of 15 amperes, the magnetic field is given by this formula as

$$\frac{2 \times 10^{-7} \times 15}{0.02} = 1.5 \times 10^{-4} \, \text{N/amp m}$$

This is stronger than the earth's field, so it would strongly deflect a compass
needle.

The discovery of electromagnetism had an immediate effect not only on
science, but also on technology. The powerful magnets that are used in steel
mills and particle accelerators are *electromagnets,* in which the magnetic field
is produced by electric currents in a coil of wire rather than in the iron atoms of
a lodestone or other permanent magnet. Perhaps the application of electromagnetism that had the greatest importance for human history was the electric
telegraph. As Ampère saw at once, the deflection of a compass needle can tell
us whether a current has been switched on in a wire even if the switch is far

away; so a message that is reduced to a series of "ons" and "offs" can be sent as far as a wire will carry a sufficiently strong current. Many prototype telegraphs were developed along these lines in the years after Oersted's discovery. One telegraph line was put in operation in 1834 by Gauss and Weber in the town of Göttingen, connecting the laboratory and the observatory. Finally, in the U.S., Samuel F. B. Morse (1791–1872) developed a practical telegraph, and in 1834 with support from Congress laid a working line between Washington and Baltimore.

Both Ampère and Oersted became scientific celebrities, and were honored by election to scientific societies throughout Europe, but their responses were quite different. Ampère combined great mathematical talent with a morose and retiring disposition—perhaps not surprising, since his father had been guillotined during the French Revolution. Many stories circulated about his absentmindedness; for instance, how he once started to do some calculations on the side of a cab that was standing in a street in Paris, and lost his work when the cab drove off. Late in life, he said that in his whole life he had had only two years of happiness.

Oersted presents an altogether more jolly picture. In the first years after his discovery of electromagnetism, he founded a society for the spread of science, and he lectured on his work in Denmark, Norway, and Germany. In 1825 he succeeded in using electric currents to isolate the element aluminum from the compound alumina. He was especially happy to be awarded the Grand Cross of the Dannebrog in 1847. He made friends with Hans Christian Andersen, who referred to Oersted as the "Great Hans Christian" and himself as the "Little Hans Christian." As the greatest Danish scientist between Brahe and Bohr, Oersted had become a national hero. In 1954, when I was a graduate student at the Niels Bohr Institute in Copenhagen, the tram that took me to work every day passed through a long and busy street whose Danish name means "H. C. Oersted's Way."

Magnetic Deflection of Cathode Rays

In Thomson's experiment the cathode ray passed through a region in which the ray was subjected to a uniform magnetic field pointing at right angles to the ray's direction. The development of the theory of magnetic forces described in the preceding section makes it possible to calculate the force this magnetic field would exert on a length of wire carrying a known electric current, but what we need to calculate here is the force the magnetic field exerts on any one of the individual particles in the cathode ray.

A simple way of inferring the magnetic force on a single charged particle from the known force on a current-carrying wire was worked out by Wilhelm Weber (1804–1890), one of the first physicists to interpret electric current as a flow of charged particles. Recall that the force on a length of wire due
to a magnetic field (at right angles to the wire) is the product of the length of
the wire, the electric current it carries, and the magnetic field. Hence, the problem is to reinterpret the product of the length of the wire and the current in
terms of the number and velocity of the individual charged particles in the
wire.

Consider a length of wire carrying an electric current. Since the distance a particle travels is just the product of its velocity and the time elapsed,
the length of the wire is equal to the product of the velocity of the particles
traveling through it and the time that it takes any of these particles to pass
from one end of the wire to the other. Multiplying this product by the electric
current, we find

$$\text{Length of wire} \times \text{Electric current} = \text{Velocity of charged particles} \times \text{Time for charge to travel length of wire} \times \text{Electric current}$$

Now look at the product of the last two factors. Since current is charge per
time, the product of the time it takes a charge to travel the length of the wire
and the electric current is just the total charge in the wire. Hence, the length
of the wire times the current equals the charge contained in the wire times the
velocity of the charged particles.* That is,

$$\text{Length of wire} \times \text{Electric current} = \text{Velocity of charged particles} \times \text{Electric charge in wire}$$

Combining the above formula with the formula on p. 48, we see that
the magnetic force on a length of wire is the product of the electric charge of
all the moving particles in the wire, their velocity, and the magnetic field. If the
particles all have the same charge and velocity, they must share equally in this

---

* Incidentally, this formula is not restricted to electric currents and charges; for instance, if a
100-kilometer-long highway carries a "current" of 1,000 automobiles per hour traveling at 50
kilometers per hour, then, since 100 km × 1,000 cars/hr = 2,000 cars × 50 km/hr, there must be
2,000 automobiles on the highway.

force. Thus, the force on any one particle due to a magnetic field at right angles to its direction will be given by

$$\begin{matrix} \text{Force on a moving particle} \\ \text{due to a magnetic field at} \\ \text{right angles to its velocity} \end{matrix} = \begin{matrix} \text{Electric charge} \\ \text{of particle} \end{matrix} \times \begin{matrix} \text{Velocity of} \\ \text{particle} \end{matrix} \times \begin{matrix} \text{Magnetic} \\ \text{field} \end{matrix}$$

For instance, the particles thrown out from the sun that strike the earth's atmosphere have electric charges of about 2×10^{-19} coulombs and velocities of about 5×10^5 meters per second, so the force that the earth's magnetic field (about 5×10^{-5} newtons/ampere meter) exerts on such a particle is roughly

$$(2 \times 10^{-19} \text{ C}) \times (5 \times 10^5 \text{ m/sec}) \times (5 \times 10^{-5} \text{ N/amp m}) = 5 \times 10^{-18} \text{ N}.$$

This is not a very big force, but these particles have a mass of roughly 5×10^{-26} kg, so their magnetic acceleration is about 5×10^{-18} N divided by 5×10^{-26} kg, or 10^8 m/sec^2—enormously greater than the acceleration of 9.8 m/sec^2 due to gravity. The force is less if the particles' velocity is not perpendicular to the magnetic field, and vanishes altogether for particles that move along the magnetic field directions. This is why the high-velocity charged particles emitted by the sun are channeled by the earth's magnetic field so that they tend to travel along the direction of the field, striking the earth near the magnetic poles and producing the beautiful northern and southern lights when they enter the atmosphere.

The shift in emphasis in Weber's work, from the question of the magnetic force on a current-carrying wire to that of the force on single particles, helped to prepare the ground for Thomson's treatment of cathode rays as streams of individual particles. In particular, Thomson was able to use the above formula for the magnetic force on a moving particle, together with the formula on p. 31, to calculate the displacement of a cathode ray due to a magnetic field at a right angle to its direction. The factor of the velocity in the numerator of the above formula for the magnetic force cancels one of the two powers of the velocity in the denominator of the formula for the displacement, so we find a displacement

$$\begin{matrix} \text{Displacement} \\ \text{of ray due} \\ \text{to magnetic field} \end{matrix} = \cfrac{ \begin{matrix} \text{Charge of} \\ \text{ray} \\ \text{particle} \end{matrix} \times \begin{matrix} \text{Magnetic} \\ \text{field} \end{matrix} \times \begin{matrix} \text{Length of} \\ \text{deflection} \\ \text{region} \end{matrix} \times \begin{matrix} \text{Length of} \\ \text{drift} \\ \text{region} \end{matrix} }{ \begin{matrix} \text{Mass of} \\ \text{ray particle} \end{matrix} \times \begin{matrix} \text{Velocity of} \\ \text{ray particle} \end{matrix} }.$$

The important point for Thomson was that, because the magnetic force is proportional to the velocity, the magnetic deflection depends on a different combination of the charge, mass, and velocity of the ray particles than does the electric deflection.

Thomson's Results

Now we will put the theory that has been developed in previous sections together with Thomson's experimental results to learn something about the cathode-ray particles. First, recall the main results we obtained above. Electric or magnetic fields at right angles to the cathode ray in the "deflection region" will produce a displacement of the ray when it hits the glass wall of the tube at the end of the "drift region," by an amount given by the formulas

$$\text{Electric deflection} = \frac{\text{Charge of ray particle} \times \text{Electric field} \times \text{Length of deflection region} \times \text{Length of drift region}}{\text{Mass of ray particle} \times \left(\text{Velocity of ray particle}\right)^2}$$

and

$$\text{Magnetic deflection} = \frac{\text{Charge of ray particle} \times \text{Magnetic field} \times \text{Length of deflection region} \times \text{Length of drift region}}{\text{Mass of ray particle} \times \text{Velocity of ray particle}}$$

Thomson knew the values of the electric and magnetic fields in the tube and the length of the deflection and drift regions, and he measured the deflections produced by the electric or magnetic forces. What, then, could he deduce about the cathode-ray particles? It is immediately clear that there was no way Thomson or anyone else could use these formulas to learn anything separately about the charge or the mass of the cathode-ray particles, since in both formulas it is only the *ratio* of these quantities that appears. Never mind—this ratio is interesting in its own right. (We will come back in Chapter 3 to the separate measurement of the electron's mass and charge.) Another problem is that neither formula could be used by itself to learn even the ratio of the charge and the mass of the cathode-ray particles, because Thomson did not know the particles' velocity. However, as has already been mentioned, this problem could be surmounted by measuring both the electric and the magnetic deflection. For

Table 2.1. Results of Thomson's experiments on electric and magnetic deflection of cathode rays.

Gas in cathode-ray tube	Material of cathode	Electric field (N/C)	Electric deflection (m)	Magnetic field $(N/amp\text{-}m)$	Magnetic deflection (m)	Deduced velocity of ray particles (m/sec)	Deduced ratio of particle mass to charge (kg/C)
Air	Aluminum	1.5×10^4	0.08	5.5×10^{-4}	0.08	2.7×10^7	1.4×10^{-11}
Air	Aluminum	1.5×10^4	0.095	5.4×10^{-4}	0.095	2.8×10^7	1.1×10^{-11}
Air	Aluminum	1.5×10^4	0.13	6.6×10^{-4}	0.13	2.2×10^7	1.2×10^{-11}
Hydrogen	Aluminum	1.5×10^4	0.09	6.3×10^{-4}	0.09	2.4×10^7	1.6×10^{-11}
Carbon dioxide	Aluminum	1.5×10^4	0.11	6.9×10^{-4}	0.11	2.2×10^7	1.6×10^{-11}
Air	Platinum	1.8×10^4	0.06	5.0×10^{-4}	0.06	3.6×10^7	1.3×10^{-11}
Air	Platinum	1.0×10^4	0.07	3.6×10^{-4}	0.07	2.8×10^7	1.0×10^{-11}

The electric deflections vary even for entries with the same electric field, because of differing cathode-ray velocities in the different cases. The magnetic deflections are the same here as the electric deflections, because in each case Thomson adjusted the magnetic field to give the same deflection as the electric field. I have calculated the results given in the last two columns from the data published by Thomson. Some of them differ by one unit in the last decimal place from the calculated values given by Thomson. I presume this is because the experimental data published by Thomson were rounded off from his actual data, and it was his actual data that Thomson used in his calculations.

instance, suppose we take the ratio of these two equations. The mass, the charge, and both lengths then cancel on the right-hand side, but the velocity does not cancel because it appears in one formula squared and in the other unsquared. This yields the simple result

$$\frac{\text{Magnetic deflection}}{\text{Electric deflection}} = \frac{\text{Magnetic field}}{\text{Electric field}} \times \text{Velocity} .$$

Since both field strengths were known and the corresponding deflections were measured, it was possible for Thomson to solve for the velocity. Then, treating the velocity as a known quantity, he could determine the ratio of charge to mass (or mass to charge) of the cathode-ray particles from either one of the formulas for the deflection of the ray, either electric or magnetic.

Now to the data. Thomson measured the electric and magnetic deflection of cathode rays for a number of different cases characterized by different values of the electric and magnetic fields, different low-pressure gases in the tube, different cathode materials, and different cathode-ray velocities. His results are shown in Table 2.1, which is adapted from his 1897 article in the *Philosophical Magazine.*[15] In all these cases, Thomson used a cathode ray in which the distance traveled by the ray while under the influence of electric and magnetic forces (the length of the deflection region) was 0.05 meters, and the distance that it subsequently traveled freely before striking the end of the tube (the length of the drift region) was 1.1 meters.

The two rightmost columns of Table 2.1 show values of the cathode-ray particle's velocity and mass/charge ratio deduced from Thomson's measurement of the electric and magnetic deflections. The formulas for calculating these quantities are worked out in Appendix B. Here, let us just check one set of results to see if they have been calculated correctly. Look at the first row in Table 2.1. For this run of the experiment, the electric and magnetic fields were 1.5×10^4 newtons per coulomb and 5.5×10^{-4} newtons per ampere-meter, the deduced value of the cathode-ray velocity was 2.7×10^7 meters per second, and the deduced ratio of particle mass to charge was 1.4×10^{-11} kilograms per coulomb (equivalent to a ratio of charge to mass of 7×10^{10} coulombs per kilogram). Using the formulas at the beginning of this section, we find the following deflections:

$$\text{Electric deflection} = \frac{(7 \times 10^{10} \text{ C/kg}) \times (1.5 \times 10^4 \text{ N/m}) \times 0.05 \text{ m} \times 1.1 \text{ m}}{(2.7 \times 10^7 \text{ m/sec})^2}$$
$$= 0.08 \text{ m,}$$
$$\text{Magnetic deflection} = \frac{(7 \times 10^{10} \text{ C/kg}) \times (5.5 \times 10^{-4} \text{ N/amp m}) \times 0.05 \text{ m} \times 1.1 \text{ m}}{2.7 \times 10^7 \text{ m/sec}}$$
$$= 0.08 \text{ m.}$$

This is in agreement with the measured deflections, which confirms that the velocity and the mass/charge ratio were calculated correctly. Incidentally, the deflection came out the same here for both electric and magnetic fields (as in the other experimental runs) for a reason of no great importance; it is just that Thomson found it convenient to adjust the magnetic field in each run until it gave the same deflection as the electric field.

The last column of Table 2.1 shows reasonable consistency. Even though the gas in the cathode-ray tube and the material of the cathode were varied from run to run, and the velocity of the cathode-ray particles varied by almost a factor of 2, the mass/charge ratios of the supposed cathode-ray particles came fairly close in all cases. This was (at least to Thomson) convincing evidence that cathode rays consisted of a single kind of particle, with a unique value of mass and charge, independent of the material of the cathode from which they were emitted.

An average of Thomson's results for the mass/charge ratio of the cathode-ray particles gives a value of 1.3×10^{-11} kilograms per coulomb. Thomson did not publish estimates of the uncertainties in his individual measurements (a failing that would cause his paper to be returned to him by any good physics journal to which it might be submitted today). However, from the spread in his values of the mass/charge ratio, we can conclude that these values must have been subject to a statistical uncertainty (in either direction) of about 0.2×10^{-11} kg/C.

Thomson's result, of a mass/charge ratio probably between 1.1×10^{-11} and 1.5×10^{-11} kg/C, can be compared with the modern value of 0.56857×10^{-11} kg/C. Evidently, Thomson did not come very close. Because his results have a fair degree of internal consistency, one suspects that there was some large systematic error in Thomson's measurements of electric and magnetic fields that pervaded all his experimental runs, but after eighty years who can tell? Thomson was not very good in handling apparatus. In fact, however, Thomson did not rely solely on his measurements of electric and magnetic deflections to determine the mass/charge ratio of the cathode-ray particles. He also employed another method, based on measurements of the heat energy deposited at the end of the tube. We will come back to this method after reviewing the concept of energy.

Flashback: Energy

Moving bodies carry the capacity to affect the objects they strike. From all our experience—of raindrops striking the earth, or bullets striking a target, or electrons striking the end of a cathode-ray tube—we know that such effects increase with the mass and the velocity of the moving body. In fact, there is a simple combination of mass and velocity that provides an extraordinarily useful measure of the capacity of moving bodies to produce effects of all sorts. It is known as the *kinetic energy,* and it is given by the formula

$$\text{Kinetic energy} = \tfrac{1}{2} \times \text{Mass} \times (\text{Velocity})^2 .$$

Energy comes in many forms, but kinetic energy is the easiest to describe and serves as the prototype for all the other forms of energy. The unit of energy in the meter-kilogram-second system is the *joule* (J). For instance, the kinetic energy of an automobile whose mass is 2×10^3 kilograms and whose velocity is 30 meters per second is

$$\tfrac{1}{2} \times (2 \times 10^3 \text{ kg}) \times (30 \text{ m/sec})^2 = 9 \times 10^5 \text{ J}.$$

The importance of this particular combination of mass and velocity arises from its relation to *work*. Work is a measure of how much is accomplished when a force is used to push something a given distance; it is simply the product of the force and the distance. We are very conscious when we lift a heavy object that the work we do is proportional both to the force that we have to exert against gravity (that is, to the object's weight) and to the height that the object is lifted. When a force is exerted on a body and is not counterbalanced by some other force, such as gravity, then the body is accelerated. In this case, the *increase in the body's kinetic energy is just equal to the work done*. (This result is proved in Appendix D.) For instance, if we push a body with a force of 1 newton through a distance of 1 meter, the increase in the body's kinetic energy is just 1 joule. This relation also works in reverse: When a moving body pushes against an obstacle, then the work that the body *does* is equal to the *decrease* in its kinetic energy. The factor of $\tfrac{1}{2}$ in the definition of kinetic energy was put in just so that the kinetic energy acquired or lost by a body would be related in this simple way to the work done on or by the body.

The relation between kinetic energy and work leads in a straightforward way to the second great property of kinetic energy: that in a broad class of circumstances it is *conserved*. If, for instance, in a game of pocket billiards the cue ball strikes the eight ball, and if the balls are not appreciably heated or otherwise changed in the collision, then—even though the cue ball will lose some kinetic energy and the eight ball will gain some—the *sum* of the kinetic energies of the two balls will be the same after the collision as it was before. This is because, according to Newton's Third Law, the force exerted by the cue ball on the eight ball is equal in magnitude and opposite in direction to the force exerted by the eight ball on the cue ball. Also, as long as the two balls are in contact, they move the same distance. Hence, the work done *on* the eight ball is equal to the work done *by* the cue ball. It follows, then, that the increase in the eight ball's kinetic energy is balanced by the decrease in the cue ball's kinetic energy, so the total kinetic energy remains conserved.

Of course, kinetic energy is not conserved when bodies act at a distance on one another—for instance, when a ball falls under the influence of the earth's gravity. In this case the falling body clearly gains kinetic energy, while the earth's kinetic energy remains essentially unchanged. This is a problem that has been encountered again and again in using the concept of energy in physics. Energy is first defined so that it is conserved in some limited context (such as billiard-ball collisions), and then in a larger context it is found not to be conserved. The response to this problem that has proved most fruitful in phys-

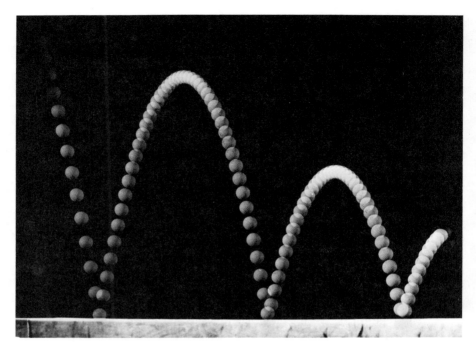

In this stroboscopic photograph of a bouncing golf ball, each exposure is separated from the preceding one by the same short interval of time. The photograph demonstrates the relation between height of fall and final velocity reached, and the interconversion of potential and kinetic energy. At the top of each bounce, the energy of the ball is all potential; at the bottom, it is all kinetic.

ics has been not to abandon the concept of energy, but to widen it—to define new kinds of energy, in such a way that the total value of all kinds of energy is conserved.

In the case of falling bodies, we can indeed define another kind of energy—the energy of position, or *potential energy*—in such a way that the total kinetic plus potential energy remains fixed. For instance, suppose we define the potential energy of a body in the gravitational field near the surface of the earth as the constant force exerted by gravity on the body times the body's height above the surface. Then the decrease in a falling body's potential energy is equal to the force exerted by gravity times the distance it falls, whether or not it falls all the way to the surface. This is just the work done by gravity, and hence it is equal to the increase in the body's kinetic energy. The decrease in potential energy is balanced by the increase in kinetic energy, so the total is conserved. As we will see, it is possible to define potential energy in a similar way for other fields of force, including electric fields.

Even when the force exerted by some field varies from place to place, we can define the potential energy of the body at a given location as the work that would be done by the field in moving the body from that location to a fixed reference point, such as the surface of the earth. The kinetic energy gained when a body moves from one location to another is just equal to the difference in its potential energy at the two locations, so the total mechanical energy, the sum of kinetic plus potential energy, remains constant.

In electric fields, as we have seen, the force on a charged body is always proportional to the body's electric charge. Thus, it is convenient in electric fields to define a quantity called *electric potential* as the potential energy of a charged particle divided by the charge. In the meter-kilogram-second system the unit of potential energy is the joule, so the unit of electric potential is the joule per coulomb, to which is given the special name of *volt*. In other words, an electric field will do one joule of work on a body carrying one coulomb of electricity when the body is moved from one location to another through an electric-potential difference of one volt. The importance of this concept arises from the fact that electric potentials can be used to characterize the environment in which electric charges move without regard to the magnitude of the charges themselves. An electric battery can be regarded as a machine for producing a fixed electric-potential difference between its positive and negative terminals or between any wires connected to them. For instance, suppose that a 1.5-volt flashlight battery drives an electric current of 0.1 ampere through a light-bulb filament, so that the electric charge transferred from one terminal of the battery to the other is 0.1 coulomb per second. The work done by the battery on each coulomb is 1.5 joules, so the rate of doing work is 0.15 joule per second (or, equivalently, 0.15 watt, since the watt is defined as a unit of power equal to one joule per second).

The idea of kinetic energy was introduced by the Dutch physicist Christian Huygens (1629–1695) in a book published posthumously in 1706. This concept, usually under the Latin name of *vis viva*, continued to be useful in the development of mechanics in the eighteenth century. In the nineteenth century its usefulness expanded greatly as the ideas of kinetic and potential energy were merged into a more general idea of energy in all its forms.

The beginning of this new and more general understanding of energy is usually credited to one of the most remarkable figures in the history of science, the American Benjamin Thompson (1753–1814), who in 1792 became Count Rumford of the Holy Roman Empire. Thompson has been described as a "loyalist, traitor, spy, cryptographer, opportunist, womanizer, philanthropist, egotistical bore, soldier of fortune, military and technical adviser, inventor, plagiarist, expert on heat, and founder of the world's greatest showplace for the popularization of science, the Royal Institution."[16] Born in Woburn, Massachusetts, he fled to England at the outbreak of the American Revolution in 1776, and thence to Germany. It was there, while serving as head of the Bavarian army, that his study of artillery led him to question existing notions of the nature of heat. It had been generally supposed that heat was a weightless fluid called "caloric," but Thompson rejected this idea on the ground that an unlimited quantity of heat can be produced by continuous mechanical work, such as the boring of a cannon barrel. Thompson concluded that heat was a form of motion, but he did not express this idea in precise terms and he did not state any equivalence between mechanical work and heat.

The next step was taken in the 1840s by Julius Mayer (1814–1878) and James Prescott Joule (1818–1889). They independently came to the conclusion that heat and mechanical energy are interconvertible; that a given amount of work always leads to the same quantity of heat and vice versa. In modern terms, the mechanical energy required to produce one calorie of heat is 4.184 joules. (The *calorie* is defined as the amount of heat needed to raise the temperature of one gram of water from 3.5° to 4.5° Celsius. It is roughly equal to the heat required to raise one gram of water one degree Celsius at any temperature.)* For example, as mentioned earlier, the force of gravity on a mass of 1 kilogram at the earth's surface is 9.8 newtons, so if the mass falls a meter its kinetic energy will be 9.8 newton meters, or 9.8 joules. If this body falls into a bucket of water, there will be a splash and the water will be set into motion, but after a while the ripples will die away and all the kinetic energy of the falling mass will have been converted into heat. The amount of heat produced in this way will be

$$\frac{9.8 \text{ J}}{4.184 \text{ J/cal}} = 2.3 \text{ cal.}$$

For instance, if the bucket contains 10 kg (10^4 g) of water, then the water temperature will increase by 2.3×10^{-4} °C. The smallness of such temperature changes explains why it took so long for the interconvertibility of mechanical energy and heat to be recognized.

Since mechanical energy and heat are interconvertible, the concept of energy can be extended to include heat. One calorie is regarded as equal to

---

* The kilogram-calorie, used to measure food energy and sometimes called the calorie in that context, is 1,000 times larger.

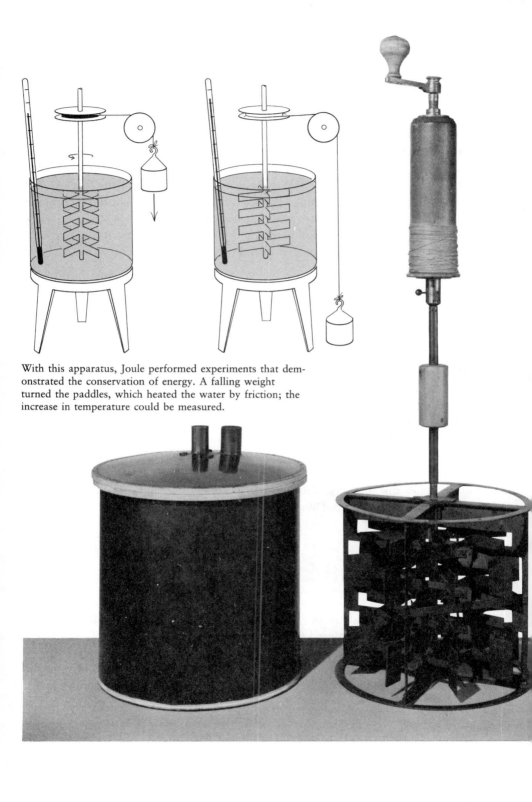

With this apparatus, Joule performed experiments that demonstrated the conservation of energy. A falling weight turned the paddles, which heated the water by friction; the increase in temperature could be measured.

4.184 joules of heat energy. When mechanical energy is turned into heat energy, as in the boring of cannon barrels, or when heat energy is turned into mechanical energy, as in a steam engine, the total energy remains conserved. The beauty of this idea is that it allows us to derive precise predictions for a great many phenomena whose nature is not entirely understood. For instance, the falling of a heavy weight into a bucket of water is a pretty complicated affair, and no one would be able to work out all the details of the splashes and ripples, but the conservation of energy can be used to predict the increase of the temperature of the water with complete confidence. It is said that Joule spent time on his honeymoon verifying the predicted increase in the temperature of water after it had passed over a waterfall.

Energy Relations in Thomson's Experiment

Now we are in a position to tie up the last loose ends in our discussion of Thomson's experiment.

First, how did Thomson know the value of the electric field between the charged aluminum plates in his cathode-ray tube? In his first five experimental runs, the electrically charged aluminum plates that produced the field were connected to a 225-volt battery. This means that the work done in carrying any electric charge from one plate to the other was 225 joules per coulomb of charge. The distance between the plates was 0.015 meters. Since work is force times distance, the electric force per coulomb times 0.015 meters was 225 joules per coulomb. Dividing by the distance, we get a force per coulomb of

$$\frac{225 \text{ J/C}}{0.015 \text{ m}} = 1.5 \times 10^4 \text{ J/C m} = 1.5 \times 10^4 \text{ N/C}.$$

(Recall that the joule is one newton-meter.) This force per coulomb is just the electric field, as entered in the first five rows of Table 2.1. (The different electric fields in the last two experimental runs were obtained with batteries of 270 volts and 150 volts instead of 225 volts.)

This little calculation suggests a different way that Thomson's experiment could have been done. If the cathode and anode are attached to the terminals of a battery or generator of known voltage, then the cathode-ray particles passing from the cathode to the anode acquire a known kinetic energy per coulomb that is just equal to this voltage.* The kinetic energy is half the mass of the particles times the square of their velocity, so dividing by the charge we have

$$\frac{\text{Voltage between}}{\text{cathode and anode}} = \frac{\frac{1}{2} \times \frac{\text{Mass of}}{\text{particles}} \times \left(\frac{\text{Velocity of}}{\text{particles}}\right)^2}{\text{Charge of particles}}.$$

Note that the combination of ray-particle parameters that appears on the right-hand side here is just the same as the combination of parameters that appears in the formula for electric deflection on p. 42, except that numerator and denominator are interchanged. Thus, in principle, the difficult measurement of deflection by electric forces could be replaced by a measurement of the voltage between the cathode and the anode.

This latter method was used in 1896–98 by Walter Kaufmann (1871–1947) of the Berlin Physics Institute to measure the mass/charge ratio of cath-

---

* There is a unit of energy that is naturally adapted to this sort of experiment: the electron volt, the energy gained or lost by an electron (or any other particle carrying the same charge) in moving through an electrical-potential difference of one volt. For instance, if the cathode and the anode of the cathode-ray tube in Thomson's or Kaufmann's experiment were connected to the negative and positive terminals of a 300-volt battery, then each electron accelerated from the cathode to the anode would pick up a kinetic energy of 300 electron volts. Unfortunately, it was not possible to relate the electron volt to ordinary units of energy like the joule or the erg without knowing the electric charge of the electron. By the definition of the volt, the work in joules is the voltage times the charge in coulombs, so the electron volt in joules just equals the electronic charge in coulombs. Since the work of Millikan (discussed in Chapter 3) we have known that the electronic charge is 1.6×10^{-19} coulombs, so the electron volt is 1.6×10^{-19} joules (more precisely, 1.602×10^{-19} joules). We could use any unit we like for elementary particle energies, but the electron volt (abbreviated eV) has become the traditional energy unit. All physicists know that the energy required to pull the electron out of the hydrogen atom is 13.6 electron volts, the energy required to pull a proton or a neutron out of a typical medium-weight nucleus is about 8 million electron volts (MeV), and so on. The cathode-ray tubes of the 1890s produced electron beams with kinetic energies of hundreds of eV. The first accelerators, developed in the 1930s by Cockcroft and Walton at the Cavendish Laboratory and by E. O. Lawrence at Berkeley, produced proton kinetic energies of the order of 10^5–10^6 eV. Energies over 10^8 eV were reached in the late 1940s, and 10^9 eV (a GeV) was attained in the 1950s. Today there are two accelerators in the world that produce proton energies over 10^{11} eV. However, no manmade accelerator matches the highest energies found in cosmic rays. These rays consist of protons and other particles that crash into our atmosphere from interstellar or perhaps intergalactic space, carrying energies up to about 10^{21} eV. Unfortunately, the high-energy cosmic rays are infrequent and interact in complicated ways with the earth's atmosphere, so they cannot substitute for manmade accelerators.

ode rays. His result for the mass/charge ratio was 0.54×10^{-11} kilograms per coulomb—quite good in comparison with the modern value of 0.5687×10^{-11} kg/C. However, as we will see in the next section, Kaufmann held back from drawing conclusions about the nature of cathode-ray particles.

Finally, we come to the method Thomson used in 1897 to obtain his most reliable value for the mass/charge ratio. In this method, the cathode ray was directed into a small metal collector that would capture the electric charge of the ray particles and would also capture their kinetic energy, converting it to heat. The ratio of the heat energy and electric charge deposited in the collector then gives the ratio of the kinetic energy and charge of *each* ray particle:

$$\frac{\text{Heat energy deposited}}{\text{Charge deposited}} = \frac{\frac{1}{2} \times \genfrac{}{}{0pt}{}{\text{Mass of}}{\text{particles}} \times \left(\genfrac{}{}{0pt}{}{\text{Velocity of}}{\text{particles}}\right)^2}{\text{Electric charge of particles}}.$$

Once again, the combination of ray parameters on the right-hand side is just the same as the combination in the formula for electric deflection on p. 42 (except for the interchange of numerator and denominator), so this combination of parameters can be determined by measuring the ratio of heat to charge deposited, rather than the deflection due to electric fields, or the voltage between cathode and anode. This is another nice example of the power of the principle of conservation of energy. Thomson had no idea at all of the detailed physical processes that occur when a cathode ray hits a metal collector, but he could be confident that the increase in heat energy of the collector had to be precisely equal to the kinetic energy lost by the cathode-ray particles when they were stopped by the collector.

Thomson's results for three different cathode-ray tubes are given in Table 2.2. The second column gives the ratio of the measured heat energy to the electric charge deposited in the collector during the time (about a second) that the cathode ray was on. The third column gives the value of the mass times the velocity divided by the charge of the cathode-ray particles, as determined according to the equation on p. 52 from the measured deflection of the cathode ray by a magnetic field. The last two columns give the values of the velocity and the mass/charge ratio of the cathode-ray particles deduced from the foregoing measured quantities. The formulas for calculating the mass-to-charge ratio and velocity are worked out in Appendix E; for now let us just check that one result comes out right. If we use the deduced values of the velocity and the

Table 2.2. Results of Thomson's experiments[17] on ratio of heat to charge deposited by cathode ray and magnetic deflection of ray.

Gas in cathode-ray tube	Measured ratio of heat energy to charge deposited (J/C)	Mass × Velocity / Electric charge (kg-m/sec-C, measured by magnetic deflection)	Deduced velocity (m/sec)	Deduced mass/charge ratio (kg/C)
Tube 1:				
Air	4.6×10^3	2.3×10^{-4}	4×10^7	0.57×10^{-11}
Air	1.8×10^4	3.5×10^{-4}	10^8	0.34×10^{-11}
Air	6.1×10^3	2.3×10^{-4}	5.4×10^7	0.43×10^{-11}
Air	2.5×10^4	4.0×10^{-4}	1.2×10^8	0.32×10^{-11}
Air	5.5×10^3	2.3×10^{-4}	4.8×10^7	0.48×10^{-11}
Air	10^4	2.85×10^{-4}	7×10^7	0.4×10^{-11}
Air	10^4	2.85×10^{-4}	7×10^7	0.4×10^{-11}
Hydrogen	6×10^4	2.05×10^{-4}	6×10^7	0.35×10^{-11}
Hydrogen	2.1×10^4	4.6×10^{-4}	9.2×10^7	0.5×10^{-11}
Carbon dioxide	8.4×10^3	2.6×10^{-4}	7.5×10^7	0.4×10^{-11}
Carbon dioxide	1.47×10^4	3.4×10^{-4}	8.5×10^7	0.4×10^{-11}
Carbon dioxide	3×10^4	4.8×10^{-4}	1.3×10^8	0.39×10^{-11}
Tube 2:				
Air	2.8×10^3	1.75×10^{-4}	3.3×10^7	0.53×10^{-11}
Air	4.4×10^3	1.95×10^{-4}	4.1×10^7	0.47×10^{-11}
Air	3.5×10^3	1.81×10^{-4}	3.8×10^7	0.47×10^{-11}
Hydrogen	2.8×10^3	1.75×10^{-4}	3.3×10^7	0.53×10^{-11}
Air	2.5×10^3	1.60×10^{-4}	3.1×10^7	0.51×10^{-11}
Carbon dioxide	2×10^3	1.48×10^{-4}	2.5×10^7	0.54×10^{-11}
Air	1.8×10^3	1.51×10^{-4}	2.3×10^7	0.63×10^{-11}
Hydrogen	2.8×10^3	1.75×10^{-4}	3.3×10^7	0.53×10^{-11}
Hydrogen	4.4×10^3	2.01×10^{-4}	4.4×10^7	0.46×10^{-11}
Air	2.5×10^3	1.76×10^{-4}	2.8×10^7	0.61×10^{-11}
Air	4.2×10^3	2×10^{-4}	4.1×10^7	0.48×10^{-11}
Tube 3:				
Air	2.5×10^3	2.2×10^{-4}	2.4×10^7	0.9×10^{-11}
Air	3.5×10^3	2.25×10^{-4}	3.2×10^7	0.7×10^{-11}
Hydrogen	3×10^3	2.5×10^{-4}	2.5×10^7	1.0×10^{-11}

mass/charge ratio given in the first row in Table 2.2, then the formula on p. 64 gives a ratio of heat energy to charge of

$$\tfrac{1}{2} \times (0.57 \times 10^{-11} \text{ kg/C}) \times (4 \times 10^7 \text{ m/sec})^2 = 4.6 \times 10^3 \text{ J/C},$$

which is indeed Thomson's measured value. (Incidentally, in this experiment the electric charge deposited in the collector was typically a few hundred-thousandths of a coulomb per second, that is, a few hundred-thousandths of an ampere, so the heat energy deposited was a few hundredths of a joule per second—enough to raise the temperature of the small collector by a few degrees Celsius per second.)

Evidently this method worked much better than the one based on the measurement of electric as well as magnetic deflection. The results for the first two cathode-ray tubes show a high degree of uniformity, and yield average values for the mass/charge ratio of 0.49×10^{-11} kilograms per coulomb—not far from the modern value of 0.5687×10^{-11} kg/C. Oddly, Thomson preferred the results obtained with his third tube, which gave a value almost two times too large. It may be that Thomson preferred the larger value of the mass/charge ratio because it agreed more closely with the result he obtained by measuring electric as well as magnetic deflection. Be that as it may, for some years Thomson made a practice of quoting the mass/charge ratio as about 10^{-11} kilograms per coulomb.

We will come back in Chapter 3 to the story of how the charge and mass of the cathode-ray particles were separately measured.

Electrons as Elementary Particles

All Thomson had done so far was to measure the mass/charge ratio of whatever particles make up the cathode rays. Yet he leaped to the conclusion that these particles are the fundamental constituents of all ordinary matter. In his own words,

> . . . we have in the cathode rays matter in a new state, a state in which the subdivision of matter is carried very much further than in the ordinary gaseous state: a state in which all matter—that is, matter derived from different sources such as hydrogen, oxygen, etc.—is of one and the same kind; this matter being the substance from which the chemical elements are built up.[17]

This was reaching very far. As Thomson recalled much later,

> *At first there were very few who believed in the existence of these bodies smaller than atoms. I was even told long afterwards by a distinguished physicist who had been present at my [1897] lecture at the Royal Institution that he thought I had been "pulling their legs."*[18]

Indeed, there was no way that the existence of smaller particles within the atom could be verified on the basis of Thomson's 1897 experiments. Thomson did not claim that he had proved it, but there were a number of hints that led Thomson toward his far-reaching conclusions.

The first of these hints was the universality of the measured ratios of mass to charge. The value of the mass/charge ratio of the cathode-ray particles did not seem to depend on any of the circumstances under which it was measured. For instance, as we saw in the preceding section, the value of this ratio was about the same for a tube containing carbon dioxide with an aluminum cathode as for a tube containing air with a platinum cathode (the fifth and sixth entries, respectively, in Table 2.1), even though the ray velocities were quite different. Thomson also quoted a result of the Dutch spectroscopist Pieter Zeeman (1865–1943) that indicated that similar values of the mass/charge ratio characterized the electric currents in atoms that are responsible for the emission and absorption of light.

(Zeeman had been studying the spectrum of the element sodium in a magnetic field. The spectrum of any element is the pattern of specific frequencies of the light that can be emitted or absorbed by atoms of that element. For instance, when a compound containing a given element is added to a flame and the light from the flame is broken up into its component colors by means of a prism or a diffraction grating, the band of colors will be found to be crossed with a number of bright lines at certain specific colors—colors corresponding to the frequencies of light being emitted by atoms of that element. The difference between light of one or another color is simply one of frequency; violet light has about twice the frequency of red light, and the other colors have intermediate frequencies. Similarly, when light from an unadulterated flame is passed through a cool vapor containing atoms of the element in question and is then broken up into its component colors, the band of colors will be crossed with a number of dark lines at precisely the same colors as the previous bright lines. These dark lines mark the frequencies at which light from the flame is being absorbed by atoms of the gas. The spectrum of sodium contains a pair of prominent lines known as the D lines, at nearby frequencies in orange light. It is these D lines that are responsible for the orange color of light from sodium

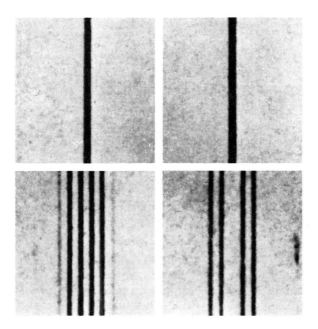

The Zeeman effect. A magnetic field splits
the spectral lines of sodium into multiple sets.

lamps, used to illuminate many highways. Zeeman observed that these D lines, which are normally quite sharp, widen in a strong magnetic field, and that the widening in frequency is proportional to the magnetic field. It was the Dutch theorist Hendrick Antoon Lorentz (1853–1928) who, in 1896, used the numerical factor in this relation of proportionality to deduce a value for the mass/charge ratio of the carriers of electric charge in atoms. It is truly remarkable that Lorentz was able to carry through this calculation a year before Thomson's discovery of the electron, fifteen years before Rutherford discovered that atoms consist of a nucleus surrounded by orbiting electrons, and seventeen years before Bohr explained how the frequencies of the light emitted or absorbed by atoms are related to the energies of the orbiting electrons. Lorentz made use of a theorem, devised by Sir Joseph Larmor, that the effect of a constant magnetic field on a system of charged particles, all of which have the same mass/charge ratio, is precisely the same as the effect that would be produced by observing the system from a coordinate system rotating at a definite frequency, now called the Larmor frequency. This frequency is proportional to the magnetic field and inversely proportional to the mass/charge ratio, but is otherwise independent of the nature of the particles, their state of motion, or the other forces that might act on them. For instance, a particle that is subjected only to magnetic forces will spiral around the lines of magnetic field at the Larmor frequency, which is just the same motion as would be seen if the particle were subject to no forces and traveled in a straight line at constant speed and if the observer's frame of reference rotated at the Larmor frequency around the direction of the magnetic-field lines. If a particle is subjected to other forces that in the absence of a magnetic field would cause it to move periodically at some natural frequency, then in the presence of a magnetic field its motion will be the superposition of three periodic motions, with frequencies equal to the natural frequency, or the natural frequency plus or minus the Larmor frequency, so the splitting in frequencies will be twice the Larmor frequency. Lorentz assumed that the frequencies of the light emitted or absorbed by atoms are equal to the frequencies of these motions, so that the splitting of the frequencies in a magnetic field would be twice the Larmor frequency for that field and hence could be used to calculate the mass/charge ratio of the carriers of electric currents in atoms. In fact, this interpretation of the frequencies at which light is emitted or absorbed by atoms is not correct, and happens to work only in certain special cases, *not* including the sodium D lines. Lorentz was lucky; although the frequencies of the two D lines of sodium are actually split by a magnetic field not into two frequencies each, but into four and six frequencies, respectively, and although the splittings among these various frequencies are not at all given by Lorentz's theory, Zeeman had not been able to resolve these separate frequencies, and by chance their overall frequency spread *is* approximately given by twice the Larmor frequency.)

Zeeman's measurements had provided a rough estimate of the mass-to-charge ratio of whatever it is that carries electric currents in atoms, and Thomson's work on cathode rays showed that these charge carriers are not just part of the architecture of the atom, but have a separate existence of their own outside as well as inside the atom. Thus it seemed that, whatever else ordinary matter might contain, it contained at least one common constituent, which could be emitted from metals as a cathode ray. The universality of these particles was soon to be verified when the so-called beta rays that were observed to be emitted by radioactive substances were found (by methods similar to Thomson's) to have the same mass/charge ratio as the cathode-ray particles. Thomson himself showed in 1899 that the negatively charged particles that are emitted in the photoelectric effect or from incandescent metal surfaces have the same mass/charge ratio as cathode rays.

The smallness of the particle mass indicated by Thomson's experiment also supported the idea that these were subatomic particles. It was already known in Thomson's time that the so-called ions that carry electric currents in solutions like salt water have various mass/charge ratios, but never a ratio less than about 10^{-8} kilograms per coulomb. (This will be discussed in some detail in the next chapter.) Thomson's result for the ratio in cathode rays was strikingly small compared with this. Of course, this might have meant either that the mass of the cathode-ray particles is smaller than the masses of ions or that their charge is greater, and for a while Thomson considered the possibility that both are true. However, it seemed more natural to suppose that ions are just ordinary atoms or molecules that become charged when they lose or gain a few units of electric charge, and if these units of charge were to be identified with the cathode-ray particles the charge of the ions would have to be comparable to the charge of the cathode-ray particles. It followed, then, that the mass of the cathode-ray particles would have to be less than the mass of the ions (and hence less than that of ordinary atoms) by a factor of about

$$\frac{10^{-11} \text{ kg/C}}{10^{-8} \text{ kg/C}} = 10^{-3}.$$

Thomson noted that this idea of very light cathode-ray particles fitted well with the observations of Phillip Lenard (1862–1947), who observed in 1894 (as Goldstein had done earlier) that cathode-ray particles could travel thousands of times farther through gases than could ordinary atoms or molecules. Since cathode-ray particles are much lighter than atoms, the possibility was open that they are the constituents of atoms.

Thomson was also predisposed to explain his observations in terms of fundamental particles by an atomic tradition, extending back to Leucippus, Democritus, and Dalton. In his 1897 paper, Thomson quoted the speculations of the English chemist William Prout (1785–1850), who in 1815 proposed that the few dozen types of atoms that were believed to make up the known chemical elements were composed of one fundamental type of atom, taken by Prout to be the atom of hydrogen. In Thomson's view, Prout was correct, but the fundamental "atom" was not the hydrogen atom but the vastly lighter cathode-ray particle. Would he have reached this conclusion if Prout and others had not made fundamental particles respectable? As we have seen, while Thomson was measuring the mass/charge ratio, a similar experiment was carried out in Berlin by Walter Kaufmann, with results that we now know were actually more accurate than Thomson's. But Kaufmann did not claim to have discovered a fundamental particle. Like Hertz and other physicists in Germany and Austria, Kaufmann was strongly influenced by the scientific philosophy of the Viennese physicist and philosopher Ernst Mach (1836–1916) and his circle, who held that it was unscientific to concern oneself with hypothetical entities like atoms that could not be directly observed. It is hard to avoid the conclusion that Thomson discovered the cathode-ray particle that we now call the electron because, unlike Mach and Kaufmann, he thought that it *was* part of the business of physics to discover fundamental particles.

At first, Thomson did not use any special name for his supposed fundamental particles. Some years earlier, the Anglo-Irish physicist and astronomer George Johnstone Stoney (1826–1911) had proposed that the unit of electricity gained or lost when atoms became electrically charged ions should be called the *electron*.[19] In the decade or so after Thomson's 1897 experiment, the reality of his fundamental particles became widely accepted and physicists everywhere began to call them electrons.

Notes

1. J. J. Thomson, "Cathode Rays," *Proceedings of the Royal Institution* 15 (1897), 419; "Cathode Rays," *Philosophical Magazine* 44 (1897), 295; "Cathode Rays," *Nature* 55 (1897), 453.

2. Plato, *Timaeus*, translated by R. G. Bury (Harvard University Press, 1929), p. 215.

3. Bede, *A History of the English Church and People*, translated by L. Sherley-Price (Penguin Books, 1955), p. 38.

4. W. Gilbert, *De magnete magnetisque corporibus, et de magno magnete telluro* (London, 1600).

5. S. Gray, "A Letter . . . Containing Several Experiments Concerning Electricity," *Philosophical Transactions of the Royal Society* 37 (1731–32), 18.

6. N. Cabeo, *Philosophia magnetica in qua magnetis natura penitus explicatur* (Ferrara, 1629).

7. C. F. Du Fay, letter to the Duke of Richmond and Lenox concerning electricity, December 27, 1733, published in English in *Philosophical Transactions of the Royal Society* (1734).

8. F. U. T. Aepinus, *Testamen theoriae electricitatus et magnetismi* (St. Petersburg, 1759).

9. B. Franklin, *Experiments and Observations on Electricity, made at Philadelphia in America* (London, 1751).

10. See, for example, A. D. Moore, ed., *Electrostatics and its Applications* (Wiley, New York, 1973).

11. Isaac Newton, *Philosophiae Naturalis Principia Mathematica*, translated by Andrew Motte, revised and annotated by F. Cajori (University of California Press, 1966).

12. Joseph Needham, *The Grand Titration: Science and Society in East and West* (Allen & Unwin, London, 1969).

13. *Epistola Petri Peregrini de Maricourt ad Sygerum de Foucaucourt, Militem, De Magnete* ("Letter on the Magnet of Peter the Pilgrim of Maricourt to Sygerus of Foucaucourt, Soldier").

14. H. C. Oersted, *Experimenta circa effectum conflictus electriciti in acum magneticum* (Experiments on the Effects of an Electrical Conflict on the Magnetic Needle), Copenhagen, July 21, 1820. For an English translation, see *Annals of Philosophy* 16 (1820), reprinted in R. Dibner, *Oersted and the Discovery of Electromagnetism* (Blaisdell, New York, 1962).

15. J. J. Thomson, "Cathode Rays," *Philosophical Magazine* 44 (1897), 295.

16. W. H. Brock, "The Man Who Played With Fire" [review of *Benjamin Thompson, Count Rumford*, by Sanborn Brown], *New Scientist*, March 27, 1980.

17. J. J. Thomson, "Cathode Rays," *Philosophical Magazine* 44 (1897), 295.

18. J. J. Thomson, *Recollections and Reflections* (G. Bell and Sons, London, 1936), p. 341.

19. G. J. Stoney, "Of the 'Electron' or Atom of Electricity," *Philosophical Magazine* 38 (1894), 418.

3

The Atomic Scale

3

The Atomic Scale

After Thomson measured the ratio of the electron's mass and charge, the great outstanding problem was to determine the mass and the charge separately. Much more was at stake here than just learning the properties of the electron, important as that was. The physicists and chemists of the nineteenth century had measured a great many other ratios of atomic properties. As we shall see in the next section, the work of John Dalton and his successors on chemical reactions had provided values for the ratios of the masses of the atoms of different elements—it was known that the atom of carbon weighs 12 times as much as that of hydrogen, that of oxygen 16 times as much as that of hydrogen, and so on. Also, as we shall see further on in this chapter, the work of Michael Faraday and others on electrolysis had yielded quite precise values for the ratios of the masses of atoms to the electric charges of ions and, by inference, to the electric charge of the electron; the ratio of the mass of the hydrogen atom to the charge of the electron was found to be 1.035×10^{-8} kilograms per coulomb. Also, atoms in solids could be presumed to be packed closely together, so measurements of the densities of solid materials provided values for the densities of atoms, that is, for the ratio of their mass and volume—for instance, gold has a density of 1.93×10^4 kilograms per cubic meter, so the ratio of the mass of a gold atom to its volume had to be somewhere in the neighborhood of 2×10^4 kilograms per cubic meter. All that was needed was one good measurement of either the charge of the electron, the mass of the electron, or the mass or volume of any single atom, and all these ratios could be instantly converted into values for the mass of the electron *and* the charge of the electron *and* the mass *and* volume of every sort of atom. In short, the scale of all atomic phenomena would then be known.

By the first years of the twentieth century there were already a number of rough methods for estimating the masses of atoms. These methods were based on a wide variety of physical phenomena: diffusion in gases, thermal radiation, the blueness of the sky, the spread of oil films, scintillations of radioactive substances, the "Brownian" motion of small particles like pollen grains produced by collisions with molecules, the effects on gas properties of the finite volume of molecules, and so on. As early as 1874, G. J. Stoney used a rough estimate based on gas properties, equivalent to 10^{-28} kilogram for the mass of a hydrogen atom, together with the mass/charge ratio of 10^{-8} kilogram per coulomb from electrolysis, to estimate that the charge of the electron is 10^{-28} kilogram divided by 10^{-8} kilogram per coulomb, or about 10^{-20} coulomb. By 1910 the precision of those measurements had improved enough (largely through Jean Perrin's work on Brownian motion) that the mass of the hydrogen atom was known to be about 1.5×10^{-27} kilogram, which yielded an electronic charge of about 1.5×10^{-19} coulomb. (Another method, based on counting radioactive disintegrations, will be discussed in Chapter 4.)

It would take us too far afield to go into these various estimates of atomic masses. In any case, the first really precise determination of these masses was based on a direct measurement of the electronic charge by the American physicist Robert Andrews Millikan (1868–1953) over the years from 1906 to 1914. Millikan was born and brought up in Iowa, and picked up his interest in physics while an undergraduate at Oberlin College. He went to Columbia University in 1893 for his Ph.D. and found himself the only graduate student there in physics. A stay in Europe was indispensable to his education, so off he went in 1895 to study at Paris, Berlin, and Göttingen. In 1896 A. A. Michelson offered him a job as a teaching assistant at the University of Chicago, then blooming under the beneficent influence of large Rockefeller donations. The position allowed Millikan to spend half his time doing research, and he accepted with alacrity. However, during the next decade he spent almost all his time on teaching and writing textbooks and very little on research. By 1906, at the age of 38, he was only just being promoted to associate professor. Almost in desperation, he took up the problem of measuring the electronic charge, and began the research that made him famous.

Recognition when it came was ample—memberships in learned academies, the presidency of the American Physical Society in 1916, and the Nobel Prize in 1923. Millikan was active in military research and development in the First World War, and then in 1921 went to the California Institute of Technology as the chairman of its executive council.

Millikan was good at raising money and publicizing worthy causes, and Cal Tech throve under his leadership, becoming what it has remained, one of America's leading centers of scientific research. Millikan also did another piece of first-rate experimental work; by careful measurement of the energies of electrons emitted in the photoelectric effect, he verified Einstein's picture of light as coming in bundles or quanta, each with energy proportional to the frequency. His later work at Cal Tech was less successful. He concerned himself much with the reconciliation of science and religion, and partly on religious grounds tried hard to prove the erroneous view that cosmic rays are electromagnetic radiation left over from the origin of matter.

For years, Millikan's measurement of the electronic charge, with the mass-to-charge ratio established by electrolysis, gave the best values for atomic masses. His method was based on work by Thomson and his collaborators at the Cavendish Laboratory, work that had yielded only a rough estimate of the electronic charge. In the last section of this chapter, we will first consider this earlier work, and then turn to the measurement by Millikan.

Flashback: Atomic Weights

Long before the existence of atoms became generally accepted, the ratios of the masses of atoms of different elements were known. The measurement of these ratios originated with the work of John Dalton (1766–1844) at the beginning of the nineteenth century. Dalton, the son of a poor Cumberland weaver, was educated at his village's Quaker school, and then worked as a schoolmaster and private tutor, moving to Manchester in 1793. The cotton mills of Manchester were at that time at the focus of the industrial revolution, and the town seems to have been filled with citizens—not generally trained at universities— who were enthusiastically following developments in science. Dalton was elected to the Manchester Literary and Philosophical Society in 1794, and began to contribute papers to it on subjects ranging from color blindness (the kind of color blindness that Dalton himself had is now called Daltonism) to gas dynamics.

The earliest record of Dalton's work on atomic weights is found in his laboratory notebooks for the years 1802–1804. Dalton observed that the weights (strictly speaking, the masses) of the various chemical elements that are required to make a given chemical compound were always in the same ratio. For instance, he found that when hydrogen was burned in oxygen to make water, 5.5 grams of oxygen were used up for each gram of hydrogen. (A warning: this is Dalton's value. The true proportion is 8 grams of oxygen to one of hydrogen. Dalton's measurements were quite poor, even by the standards of his time.) This is not at all like ordinary cookery. In baking a cake, one can always use a little more or a little less butter for each pound of flour, but one still gets a cake—perhaps a little too buttery or a little too dry, but still a cake. In contrast, if one uses a little more or a little less than 8 grams of oxygen for each gram of hydrogen, one does not get water that is a little

Robert A. Millikan with cosmic-ray apparatus.

oxygen-rich or oxygen-poor; one gets ordinary water, with a little oxygen or a little hydrogen left over.

The most important part of Dalton's work was not his rather inaccurate measurements, but rather his interpretation of them in terms of atoms. Dalton reasoned that if water consisted of particles (later to be called *molecules*) each of which contained one atom of hydrogen and one atom of oxygen, then the recipe of 5.5 grams of oxygen to each gram of hydrogen for water could be explained if the atom of oxygen weighed 5.5 times as much as the atom of hydrogen. In this manner, Dalton worked out the atomic weights shown in Table 3.1. In Dalton's meaning, *atomic weight* means the weight, or mass, of an atom relative to that of hydrogen. Dalton, of course, had no idea what the weight of any atom might be in ordinary units like pounds or kilograms.

The atomic weights in Dalton's table are in fact all wrong, partly because of Dalton's errors in measurement but mainly because Dalton did not know the correct proportions of atoms in the molecules of chemical compounds. For instance, Dalton assumed that the water molecule consists of one atom of oxygen and one atom of hydrogen, but today everyone knows that the correct formula for water is H_2O—that is, *two* atoms of hydrogen and one atom of oxygen in each molecule. (The subscripts give the number of atoms of each element in the molecule, a 1 being understood where no subscript appears.) Dalton's measurement that 5.5 grams of oxygen were consumed for each gram of hydrogen in making water means, then, that one atom of oxygen weighs 5.5 times as much as two atoms of hydrogen, or 11 times as much as one atom of hydrogen. This is closer to the true atomic weight of oxygen, now known to be approximately 16. Table 3.2 gives the chemical formulas for various compounds Dalton used in preparing his table of atomic weights, together with the true formulas. Table 3.3 gives the precise modern atomic weights, together with the values found by Dalton and the values he would have found if he had known the correct chemical formulas listed in Table 3.2.

Table 3.1. Dalton's 1803 values of atomic weights.

Element	Atomic weight
Hydrogen	1 (by definition)
Nitrogen ("Azot")	4.2
Carbon ("Carbone")	4.3
Oxygen	5.5
Sulfur	14.4

ELEMENTS

		W.t				W.t
⊙	Hydrogen.	1	⊕	Strontian	46
⊖	Azote	5	✳	Barytes	68
⬤	Carbon	54	Ⓘ	Iron	50
○	Oxygen	7	Ⓩ	Zinc	56
◉	Phosphorus	9	Ⓒ	Copper	56
⊕	Sulphur	13	Ⓛ	Lead	90
◉	Magnesia	20	Ⓢ	Silver	190
◉	Lime	24	ⓖ	Gold	190
⬤	Soda	28	Ⓟ	Platina	190
⬤	Potash	42	⊛	Mercury	167

Dalton's symbols for chemical elements. Some of these are now known to be compounds, not elements.

Table 3.2. Chemical formulas for various compounds, as used by Dalton and as known today.

Compound	Dalton's formula	True formula
Water	HO	H_2O
Carbon dioxide ("carbonic acid")	CO_2	CO_2
Ammonia	NH	NH_3
Sulfuric acid	SO_2	H_2SO_4

C is carbon, H hydrogen, N nitrogen, O oxygen, and S sulfur.

Table 3.3. Modern values of atomic weights for five elements, together with the values obtained by Dalton and the values he would have found by using the correct chemical formulas.

Element	Modern atomic weight	Dalton's (1803) atomic weights	Dalton's atomic weights using correct chemical formulas
Hydrogen	1.0080	1	1
Carbon	12.0111	4.3	8.6
Nitrogen	14.0067	4.2	12.6
Oxygen	15.9994	5.5	11
Sulfur	32.06	14.4	57.6

The modern atomic weights given here are the weights relative to 1/12 the weight of the atom of carbon (or, more precisely, of the most common isotope of carbon, ^{12}C), but this is pretty nearly the same as the weight of the hydrogen atom. If these weights were taken relative to hydrogen, they would be 0.8 percent smaller.

 The correct formulas for chemical compounds were deduced through a further development of the atomic theory. On December 31, 1808, Joseph Louis Gay-Lussac (1778–1850), professor at the Sorbonne, read a memoir to the Societé-Philomathique in which he reported that, although all elements combine in definite proportions of weights, gases also combine in definite proportions of *volumes*. For instance, two volumes of hydrogen plus one volume of oxygen yields two volumes of water vapor; one volume of nitrogen plus three volumes of hydrogen yields two volumes of ammonia, and so on (with "volume" here understood to mean any unit of volume—a liter, a half-liter, a cubic mile, or what have you).

The explanation of the law of combining volumes was offered in 1811 by Amadeo Avogadro, Conte di Quaregna (1776–1856), professor of physics at the University of Turin. Avogadro hypothesized that *equal volumes of any gas at a given temperature and pressure always contain equal numbers of the particles of the gas,* which Avogadro called *molecules.* For instance, the fact that two liters of hydrogen always combine with one liter of oxygen (at the same temperature and pressure) in forming water suggests immediately that the water molecule contains twice as many atoms of hydrogen as of oxygen; this is how we know that water is H_2O. There is an apparent difficulty here: If every water molecule contains one oxygen and two hydrogen atoms, then why does one liter of oxygen and two liters of hydrogen yield two liters of water vapor, and not just one liter? The answer, Avogadro realized, is that under ordinary conditions the molecules of oxygen and hydrogen contain two atoms each (called "molecules elementaires" by Avogadro), not just one. This doubles the number of atoms of hydrogen and oxygen per liter, and hence doubles the number of water molecules and the volume of water vapor produced from given volumes of hydrogen and oxygen. On this basis, the chemical reactions for the production of water and ammonia read $2H_2 + O_2 \rightarrow 2H_2O$ and $N_2 + 3H_2 \rightarrow 2NH_3$. The number in front of the chemical symbol for each molecule shows how many molecules of that chemical compound participate in the reaction; hence, according to Avogadro's hypothesis, these numbers also give the relative volumes of the gases needed in the reactions.

Avogadro's hypothesis was a brilliant guess. Today we understand it in terms of the kinetic theory of gases: The pressure that a gas exerts on a wall is given to a good approximation by the product of the temperature, the number of gas molecules per liter, and a universal constant called Boltzmann's constant, irrespective of the nature of the gas molecules. (See Appendix F.) Hence, for a given temperature and pressure, we always have the same number of molecules per liter. In Avogadro's day his hypothesis had to be justified purely empirically, by showing that it works. That is, adopting Avogadro's hypothesis, one could work out the chemical formulas for different gaseous compounds, just as we could conclude above that water is H_2O. Then, from the ratios of the weights of the elements and compounds that participate in various reactions, one could determine the atomic weights (relative to, say, hydrogen), just as done by Dalton. The check here is that the atomic weight of a given element must come out the same in all reactions. If Avogadro's hypothesis were wrong, it would lead to the wrong chemical formulas for various compounds, and hence to inconsistent atomic weights for different reactions.

A word on terminology: The *molecular weight* of a chemical compound is equal to the sum of the atomic weights of the atoms that make up a molecule of the compound. For instance, the molecular weight of a water molecule is $2 + 16 = 18$. For an element like helium, whose molecules consist of a single atom, the molecular weight is the same as the atomic weight. Some molecules, such as DNA, have molecular weights in the millions. Chemists often use as a unit of mass the *mole,* defined as the number of grams equal to the molecular weight—one mole of hydrogen gas is 2 grams, one mole of water is 18 grams, and so on. The mole is a useful unit because a mole of any substance always contains the same number of molecules—the heavier the molecules, the more grams per mole. This number, the number of molecules per mole, is known as *Avogadro's number.* Avogadro, alas, had no way of calculating Avogadro's number; that had to wait until developments discussed below.

Avogadro used chemical formulas deduced on the basis of his hypothesis to determine a number of atomic weights with pretty fair accuracy. The work was then taken up by others, especially Jöns Jakob Berzelius (1779–1848), professor of chemistry at the University of Stockholm. Berzelius published tables in 1814, 1818, and 1826 that provided quite good values of the atomic weights of many elements. By the end of the nineteenth century, although not all physicists and chemists believed in the existence of atoms, they were accustomed to use tables of atomic weights as a tool in their everyday work.

Even for those nineteenth-century physicists who did believe in the reality of atoms, there remained one great uncertainty in the interpretation of atomic weights. When we say that a certain element has a given atomic weight, is this the weight of all atoms of the element (relative, say, to hydrogen) or merely the average weight of such atoms? One of the early workers on gas discharges, Sir William Crookes, guessed in 1886 that the atomic weights measured by chemists were in fact averages of the weights of different atoms of the same element. We now know that this is true. Almost all elements come in different forms, called *isotopes.* The atoms of different isotopes of the same element are almost indistinguishable chemically, but have different atomic weights.

The story of the discovery of isotopes takes us well into the physics of the twentieth century. Nevertheless, even though this is a "flashback" section, no discussion of atomic weights would be complete without some account of how our modern understanding of isotopes developed.

Soon after the discovery of radioactivity in 1897 it was found that there are different forms of certain chemical elements, identical in chemical behavior but very different in radioactive behavior. For instance, lead is generally nonradioactive, but lead that is associated with uranium-bearing minerals exhibits radioactivity of its own, and this radioactivity persists even when all elements that can be chemically separated from lead have been removed. It soon became clear that these varieties of an element with different radioactive behavior consist of atoms of different atomic weight. In 1910 Frederick Soddy called the varieties of the same element *isotopes,* because they were in the same place (*iso* means same; *tope* means place) in the list of chemical elements. However, radioactivity was still somewhat mysterious, and it seemed possible that the occurrence of isotopes was a peculiarity of heavy radioactive elements.

The discovery that ordinary, nonradioactive light elements also have isotopes was due to J. J. Thomson. Not surprisingly, the technique he used was based on the electric and magnetic deflection of rays produced in a cathode-ray tube—not the usual cathode rays, which consist of electrons, but rays consisting of heavy positive particles. In 1886 Eugen Goldstein, who had given cathode rays their name, noticed that, in a cathode-ray tube with a hole pierced in the cathode, a ray would emerge from the hole and travel in the direction *away* from the anode, producing a visible line of light in the rarefied gases within the tube. He called these rays *canal rays (Kanalstrahlen)*. Wilhelm Wien (1864–1928) succeeded in 1897 in deflecting the canal rays with electric and magnetic fields, and from the direction and the amount of this deflection he concluded that they consisted of positively charged particles with mass/charge ratios thousands of times that of Thomson's cathode rays and comparable to the mass/charge ratios of electrically charged atoms, as measured in electrolysis (discussed in the next section). He concluded that the particles of these canal rays were atoms or molecules of the gas within the tube that were given a positive charge when electrons were knocked out of them by the cathode ray as it traveled from cathode to anode, and were then attracted to the negatively charged cathode and repelled by the positively charged anode. Most of these accelerated positively charged particles (or ions) just hit the cathode, but a certain fraction happen to pass through the hole in the cathode and emerge on the other side as canal rays.

The study of canal rays was difficult because some canal-ray particles, after passing through the hole in the cathode, would strike a gas molecule and gain or lose an extra electron. The mass/charge ratios measured by Wien were actually averages of their values before and after these sudden changes in electric charge. Thomson solved this problem by using a tube in which the gas pressure on the side of the cathode opposite the anode could be kept very low, so that the probability of collisions between ray particles and gas molecules was minimal. He could then measure the mass-to-charge ratios of different positively charged atoms and molecules with fair precision.

In 1913 Thomson observed that the canal rays formed in neon gas exhibit two different values for the mass/charge ratio, one 20 times and the

Francis Aston with his mass spectrograph in the Cavendish Laboratory, Cambridge.

other 22 times that of singly charged hydrogen atoms. The electric charges were all the same, so Thomson concluded that there are two distinct isotopes of neon, one with atomic weight 20 and the other with atomic weight 22. The atomic weight of neon had previously been measured to be 20.2. Because this is an average atomic weight, it meant that ordinary neon (found in our atmosphere) is a mixture of these two isotopes, with 10 percent of all neon atoms in the form of the heavier isotope, ^{22}Ne, and 90 percent in the form of ^{20}Ne. (Note that 90 percent of 20 plus 10 percent of 22 equals the observed atomic weight, 20.2.) Neither isotope of neon is radioactive, and the occurrence of isotopes was thus shown to be independent of the presence of radioactivity.

Thomson's work was continued after World War I by another physicist at the Cavendish Laboratory, Francis William Aston (1877–1945), who had been Thomson's assistant before the war. Aston incorporated the method, familiar by then, of deflecting rays by electric and magnetic fields in a greatly improved new device called a mass spectrograph. With it he was able not only to confirm Thomson's result concerning neon isotopes, but also to discover a whole host of new isotopes, including two of chlorine (^{35}Cl and ^{37}Cl), three of silicon (^{28}Si, ^{29}Si, and ^{30}Si), three of sulfur (^{32}S, ^{33}S, and ^{34}S), and a third of neon (^{21}Ne). In fact, most of the lighter elements have several nonradioactive isotopes.

Aston's precise measurements of isotopic atomic weights revealed a striking common feature, which Aston stated in 1919 as a whole-number rule: If atomic weights are expressed relative to 1/16 the weight of ^{16}O (or, as is done nowadays, 1/12 the weight of ^{12}C), then all atomic weights of pure isotopes are very close to whole numbers. This had been noticed as a rough rule of thumb soon after Dalton's work, and in 1815 William Prout (1785–1850) had drawn the natural conclusion that the atoms of all chemical elements are made up of whole numbers of some one fundamental particle, which Prout guessed to be the atom of hydrogen. However, for a long time there seemed to be an obstacle to this idea in the fact that the atomic weights of some elements were not at all close to whole numbers. The notorious example was chlorine, with atomic weight 35.45. Aston was able to show that this atomic weight was really an average of the atomic weights of two isotopes of chlorine, ^{35}Cl and ^{37}Cl, with atomic weights very close to 35 and 37 and with abundances of 77.5 percent and 22.5 percent respectively. Table 3.4 lists up-to-date values of the atomic weights of some isotopes of a few common elements. Evidently, Prout's

Table 3.4. Atomic weights of some isotopes of a few representative elements.

Element	Isotope	Atomic weight
Hydrogen	^{1}H	1.007825
	^{2}H	2.01410
Helium	^{4}He	4.0026
Carbon	^{12}C	12 (by definition)
	^{13}C	13.00335
Oxygen	^{16}O	15.99491
	^{17}O	16.9991
Neon	^{20}Ne	19.99244
	^{21}Ne	20.99395
	^{22}Ne	21.99138
Chlorine	^{35}Cl	34.96885
	^{37}Cl	36.9659
Uranium	^{235}U	235.0439
	^{238}U	238.0508

hypothesis and Aston's whole-number rule work very well indeed, especially for atoms of medium atomic weight.

Today we know that the occurrence of isotopes is due to the fact that the nuclei of atoms are composed of electrically neutral particles—neutrons—as well as positively charged protons. It is the number of protons in the nucleus that determines the number of electrons in the atom, which have to neutralize the protons' charge. Therefore the chemical nature of an element is at bottom a matter of the number of protons in its nuclei. All hydrogen atoms have one proton in the nucleus, all helium atoms two protons, and so on, up to lawrencium with 103 protons. The atoms of isotopes of the same element all have the same number of protons and electrons, but they have different numbers of neutrons and hence different atomic weights. Neutrons and protons have nearly the same mass (about the mass of a ^{1}H atom), and electrons are much lighter; thus, the atomic weight of an isotope is very nearly equal to the total number of protons and neutrons contained in its atomic nuclei, which of course is a whole number. However, none of this could be known without further developments in nuclear physics. After we look at these developments in Chapter 4, we will also be able to understand the implications of the small departures from Aston's whole-number rule—departures that are as important as the rule itself.

A postscript: Since different isotopes of the same element are chemically almost indistinguishable, they cannot be separated by ordinary chemical means. Just before World War I, Aston had developed a method of separation that relied on the faster diffusion of light atoms through porous materials like pipe clay. After allowing a sample of neon gas to diffuse many times through such materials, he found that it had been slightly enriched in the lighter isotope, ^{20}Ne. However, the first nearly complete separation of one isotope of an element from the others did not come until 1932, when Harold Urey (1893–1981) and others succeeded in preparing nearly pure samples of heavy water, the oxide of ^{2}H.

During World War II the United States urgently needed to separate the uranium isotope ^{235}U, which can be used to make nuclear weapons, from the more common isotope ^{238}U. The methods adopted by the Manhattan Project were precisely those developed at the Cavendish Laboratory: the electromagnetic deflection method of Wien, Thomson, and Aston, and the gaseous-diffusion method of Aston. The gaseous-diffusion method in the end proved more feasible, and provided the ^{235}U exploded at Hiroshima. (The Nagasaki bomb employed a different element, plutonium.) There are now easier methods, and we face the frightening prospect of living in a world in which ^{235}U as well as plutonium can be obtained all too easily by many governments.

Flashback: Electrolysis

One other quantitative property of atoms that is important to our story was measured in the first part of the nineteenth century, long before the discovery of electrons or the atomic nucleus. Strictly speaking, this discovery concerns not just atoms but also *ions,* the electrically charged molecules that carry electric currents in most conducting liquids. This property is the ratio of atomic masses to ionic charges, and it is measured not by deflecting an electric current with electric or magnetic fields, as Thomson did, but by simply weighing the material produced in the electrochemical process known as electrolysis.

Electrolysis was discovered more or less by accident in April 1800 by William Nicholson (1753–1815) and Anthony Carlisle (1768–1840). While studying the operation of electric batteries, they put a drop of water at the connection between a wire and the battery to improve the electrical contact. They noticed that bubbles of gas were produced where the wire entered the water. When they immersed wires from the terminals of a battery in a tube of water to study the phenomenon in more detail, they observed that hydrogen gas was produced at the wire attached to the negative terminal and oxygen gas at the positive wire. It was soon found that other substances could be chemically decomposed in this way. The most extensive experiments were those of Sir Humphrey Davy (1778–1829), professor of chemistry at Rumford's recently founded Royal Institution. Davy found that various salts could be decomposed by passing an electric current through molten salts or solutions of the salts in water, often with a metal appearing as a plating and a gas as bubbles on the immersed conductors (called electrodes) attached respectively to the negative and positive terminals of the battery. For instance, in the electrolysis of molten table salt, the metal sodium appears at the negative electrode and the gas chlorine at the positive electrode. It was through these experiments on electrolysis that Davy discovered the elements sodium and potassium, which although present in many common compounds are so chemically reactive that they are never found as free elements.

It took some time to develop a detailed understanding of these phenomena—in part because the chemists of the early nineteenth century knew little about atoms or molecules, and nothing at all about electrons, and also because the process of electrolysis is quite complicated. A substantially correct theory was finally worked out in the 1830s by Michael Faraday (1791–1867). Faraday began as a journeyman bookbinder, and educated himself by reading the books he bound. Seeking laboratory employment, Faraday impressed Davy in an interview and was hired in 1812 as an assistant to work on chemical experiments. In 1831 he succeeded Davy as director of the Royal Institution's laboratories and began his work on electricity. We saw in Chapter 2 the usefulness of Faraday's concept of lines of electric field and it was Faraday who discovered the phenomenon of induction, by which changes in magnetic fields produce electric fields.

Here in brief is the modern understanding of electrolysis, essentially as worked out by Faraday: A certain fraction of the electrically neutral molecules of a liquid, such as water, are normally dissociated into positively and negatively charged submolecules, for which Faraday introduced the name *ions*.* For instance, under ordinary conditions about 1.8×10^{-9} of the molecules in pure water are dissociated (for complicated reasons) into a positive hydrogen ion, H^+, and a negative hydroxyl ion, OH^-. We have known since the discovery of the electron that positive ions such as H^+ are just molecules (in this case, a single atom) that have lost one or more electrons (for H^+, just one), and that negatively charged ions such as OH^- are molecules that have gained one or more electrons; however, this information was not needed in Faraday's theory.

Now, suppose that conductors attached to the positive and negative terminals of an electric battery (for which Faraday introduced the name *electrodes*) are immersed in the liquid. The positive ions in the immediate vicinity of the negative electrode will be attracted to it. On contact they will take up a negative electric charge from the battery (a charge that we know today is borne by electrons), and will materialize as neutral molecules. For instance, in the electrolysis of water, this reaction is $2H^+ + 2e^- \rightarrow H_2$. Here two electrons and two hydrogen ions are participating, because, as Avogadro discovered, the normal hydrogen molecule consists of two atoms of hydrogen. Similarly, at the positive electrode, negative ions will give up their negative charge (electrons) to the battery, and will also appear as ordinary molecules. In the electrolysis of water, this reaction is $4OH^- \rightarrow 2H_2O + O_2 + 4e^-$, with the oxygen, like the hydrogen, appearing as bubbles at the electrodes. These reactions create a deficiency of positive ions at the negative electrode and a deficiency of negative ions at the positive electrode, so new ions will be attracted to the vicinity of the electrodes and the process will continue. The negative charge that is given up to the battery at the positive terminal and taken back from it at the negative terminal flows through the wires and the battery as an ordinary electrical current, whose strength can easily be measured (for instance, as in an ordinary ammeter, by measuring the magnetic force it produces).

---

* Faraday did introduce the terms *ion* and *electrode*, as well as *anion* and *cation* for positive and negative ions and *anode* and *cathode* for positive and negative electrodes, but he did not invent these terms. They were constructed at Faraday's request from Greek roots by the Master of Trinity College, Cambridge, Dr. William Whewell, and then used by Faraday in his writings.

Faraday's electrolysis equipment.

The same picture applies to the electrolysis of other materials. For instance, in the electrolysis of silver chloride the molecule AgCl breaks up into ions Ag^+ and Cl^- (Ag is silver, Cl chlorine); the reactions at the negative and positive electrode are respectively $Ag^+ + e^- \rightarrow Ag$ and $2Cl^- \rightarrow 2e^- + Cl_2$. The chlorine appears as a gas whose molecules contain two atoms each, and the silver as a monatomic plating on the negative electrode.

In all these reactions, definite numbers of molecules of each type are produced when a given amount of electric charge flows through the wires and the battery. Suppose for convenience that we take as the unit of charge the amount needed to produce one silver atom in the electrolysis of silver chloride. Then to produce each chlorine molecule one needs two units of charge, and for each hydrogen and each oxygen molecule (not atom) produced in the electrolysis of water one needs two and four units, respectively. The electrolytic unit of electricity used here is now known to be the charge of an electron. To Faraday it was just a certain irreducible quantity of charge, multiples of which are transferred between ions and electrodes in electrolysis. It was in this sense, as the basic unit of electricity in electrolysis, that Stoney in 1874 introduced the term *electron*.

This picture of electrolysis was suggested to Faraday by his measurement of the relative amounts of various materials produced. For instance, in the electrolysis of water any electric current will always produce a mass of oxygen eight times as large as the mass of hydrogen. This is what we should expect on the basis of Faraday's theory: The production of each oxygen molecule requires four units of electricity, while the production of each hydrogen molecule requires only two units; hence a given current will produce oxygen molecules at a rate half that for hydrogen molecules. However, as we saw in the previous section, the mass of each oxygen molecule is 16 times that of a hydrogen molecule, so for each gram of hydrogen one produces $\frac{1}{2} \times 16 = 8$ grams of oxygen.

Faraday had no way of knowing the magnitude of his electrolytic unit of electric charge in ordinary units such as coulombs, just as Dalton could not know the magnitude of his unit of atomic weight in ordinary units such as grams. However, the ratio of these units could now be determined easily. By weighing the amount of silver deposited on the negative electrodes in the electrolysis of salts like silver chloride, it was found that a one-ampere current flowing for one second would produce about 10^{-6} kilograms of silver, and proportionally more for stronger currents or longer times. One silver atom is produced for each unit of charge, so the number of silver atoms in the 10^{-6} kilograms of silver must equal the number of units of charge transferred by one ampere in one second, which is defined as one coulomb of charge. It follows then that the ratio of the mass of the silver atom to the unit of electric charge is about 10^{-6} kilograms per coulomb. The atomic weight of silver is about 108 times that of hydrogen, so the ratio of the mass of the hydrogen atom to the unit of electric charge is smaller than that for silver by a factor of 108, or about 10^{-8} kilograms per coulomb.

This is usually expressed in somewhat different terms. Since a mole of any substance always contains the same number of molecules (see p. 82), the amount of electric charge required to produce one mole of any substance is just equal to the number of units of electricity required per molecule (one for silver, two for hydrogen and chlorine, four for oxygen) times a universal constant known as the *Faraday*. The Faraday, which is the electric charge per electrolytic unit of electricity times Avogadro's number, the number of molecules per mole, was known with some precision by the end of the nineteenth century as 96,850 coulombs per mole. The hydrogen atom has an atomic weight of 1.008, so one mole of hydrogen atoms is 1.008 grams, or 1.008×10^{-3} kilograms. Thus, the ratio of the mass of a hydrogen atom to the unit of electric charge was known as

$$1.008 \times 10^{-3}/96,580 = 1.044 \times 10^{-8} \text{ kg/C}.$$

After Thomson's discovery of the electron, it was natural to identify the electrolytic unit of electric charge as simply the charge of the electron. On this basis, the ratio of the mass of the hydrogen atom to the charge of the electron was known to be 1.044×10^{-8} kilograms per coulomb. It was this information gleaned from electrolysis, together with Thomson's measurement of a mass/charge ratio of about 10^{-11} kilograms per coulomb for electrons, that led Thomson to conclude that atoms are thousands of times heavier than the electrons they contain.

Measuring the Electronic Charge

The charge of the electron was first measured in a series of experiments at the Cavendish Laboratory by Thomson and his colleagues J. S. E. Townsend (1868–1957) and H. A. Wilson (1874–1964). Their methods were based on the fact, discovered shortly before at the Cavendish by Thomson's student Charles Thomson Rees Wilson (1869–1959), that ions could serve to start the growth of droplets of water in humid air—a role usually played by grains of dust. Wilson's work led to the development of the cloud chamber, in which

C. T. R. Wilson's cloud chamber, which makes the trails of ionizing particles visible.

moving charged particles produce visible tracks of water droplets when a humid atmosphere is suddenly expanded. The cloud chamber did much to convince everyone of the reality of subatomic particles. However, what concerns us now is the fact that water droplets can form even around single ions, so that a measurement of the charge/mass ratio of these droplets together with a separate measurement of their size could yield a value for the charge of an ion, and by inference a value for the charge of the electron.

Townsend's method used the ions that are naturally present in the gases produced by electrolysis. The droplets of water that form around these ions were much too small for their size to be measured directly, so Townsend used a method based on the speed of falling droplets (a method that was to be repeated in most future measurements of the electron's charge). Under the influence of gravity, a droplet of water will accelerate until the viscous drag of the air just cancels the force of gravity, and will then fall at a constant speed. The force of gravity on the droplet is equal, according to Newton's Second Law, to the mass of the droplet times the acceleration of 9.8 meters per second per second with which objects would fall if not supported by other forces:

$$\text{Gravity force on droplet} = \text{Mass of droplet} \times 9.8 \text{ m/sec}^2.$$

The viscous drag of the air, on the other hand, depends on both the radius of the droplet and its velocity. From the theoretical work of Sir George Stokes (1819–1903) in 1851, it was known that this force is given by the formula

$$\text{Drag force on droplet} = 6\pi\eta \times \text{ Radius of droplet} \times \text{Velocity of droplet},$$

where η is a numerical quantity that gives the viscosity or "stickiness" of the air, and is known through various measurements (for instance, of the rate of falling of larger bodies of known size) to be about 1.82×10^{-5} newton-sec/m^2. Now of course the drag force acts in a direction to oppose the droplet's motion, so if the velocity reaches a certain value the two will cancel. This is what happens: The droplet accelerates under the influence of gravity until its velocity approaches the point where drag cancels gravity, and falls at that velocity from then on. At the velocity of steady fall, the right-hand sides of the two above equations must be equal:

$$\text{Mass of droplet} \times 9.8 \text{ m/sec}^2 = 6\pi\eta \times \text{Radius of droplet} \times \text{Velocity of steady fall.}$$

By measuring the velocity of the droplet, Townsend therefore obtained one relation between the mass and radius of the droplet. Another relation is provided by the fact that the mass of the droplet must equal its volume times the known density (10^3 kilograms per cubic meter) of water. Using the familiar formula for the volume of a sphere, this gives

$$\text{Mass of droplet} = \frac{4\pi}{3} \times (\text{Radius of droplet})^3 \times \text{Density of water.}$$

We now have two equations relating two unknowns, the mass and radius of the droplet, so it is easy to solve for both. (This is done in Appendix G.) In this way, Townsend was able to calculate the average mass of the droplets in his falling cloud of water vapor.

The cloud of droplets was then passed through sulfuric acid, which absorbed the water, and Townsend measured the electric charge picked up by the acid and the increase in its weight due to the absorbed water. Taking the ratio gave the charge/mass ratio of the droplets, and multiplying by the previously determined mass of each droplet gave the charge on each droplet. Townsend's 1897 results were that the charges were 0.9×10^{-19} coulombs for positive ions and 1.0×10^{-19} coulombs for negative ions, with the 10 percent discrepancy easily accounted for by experimental uncertainties.

In Thomson's method of measuring the electronic charge, the ions were produced by exposing air to x rays. Instead of absorbing the water droplets in sulfuric acid, he measured their total mass and electric charge by very indirect means involving measurements of the electrical conductivity of the air and the temperature change during the expansion that produced the water droplets. The size of the individual droplets was measured (as in Townsend's experiment) by measuring the rate at which the cloud fell. Thomson's 1898 result was that the ionic charge was about 2×10^{-19} coulombs. With improvements in his technique, by 1901 he was quoting a value of 1.1×10^{-19} coulombs.

H. A. Wilson's method, like Thomson's, used ions produced by x rays, but the resulting cloud of water droplets was subjected to a strong vertical electric field. With the field off, the size and mass of the droplets could be measured from the rate of fall of the cloud, as in Townsend's and Thomson's experiments. With the field on, the droplets were subject to three forces: the force of gravity (which depends on the previously measured droplet mass), the viscous drag of the air (which depends on the previously measured radius of the droplets and their observed velocity), and the electric force on the droplet (the product of the electric charge of the droplet and the electric field). From the condition that the velocity has reached a steady value at which all three forces are in balance, one can then solve for the only unknown quantity, the electric charge carried by the droplets. (This calculation is also carried out in Appendix G.) In 1903 Wilson reported a charge of 1.03×10^{-19} coulombs.

These results were in reasonable agreement with each other, but they were not regarded as being very precise, and in fact they were not. (As we will see, the true charge of the electron is 60% larger.) Nevertheless, the evidence for the atomicity of electricity was good enough to convince many who, like Mach, had been skeptical about the reality of atoms. I. B. Cohen and G. Holton both quote the admission by Wilhelm Ostwald (1853–1932), who had been a leading opponent of atomism, in the 1908 edition of Ostwald's *Outlines of General Chemistry:* "I am now convinced that we have recently

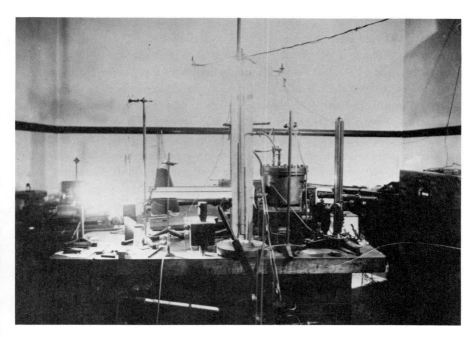

The apparatus used by Millikan in his oil-drop experiments.

become possessed of experimental evidence of the discrete or grained nature of matter, which the atomic hypothesis sought in vain for hundreds and thousands of years." The evidence quoted by Ostwald was Perrin's experiments on Brownian motion and Thomson's measurement of the electronic charge.

Now we come to Millikan. It was some time around 1906 that Millikan began an effort to measure the charge of the electron more accurately than had been possible at the Cavendish. At first he simply repeated the method of H. A. Wilson, but soon he introduced a crucial improvement.* Instead of droplets of water that condensed from a humid atmosphere, he used droplets of mineral

---

* After this was written there appeared a remarkable posthumous memoir that throws some doubt on Millikan's leading role in these experiments. Harvey Fletcher (1884–1981), who was a graduate student at the University of Chicago, at Millikan's suggestion worked on the measurement of electronic charge for his doctoral thesis, and co-authored some of the early papers on this subject with Millikan. Fletcher left a manuscript with a friend with instructions that it be published after his death; the manuscript was published in *Physics Today*, June 1982, page 43. In it, Fletcher claims that he was the first to do the experiment with oil drops, was the first to measure charges on single droplets, and may have been the first to suggest the use of oil. According to Fletcher, he had expected to be co-author with Millikan on the crucial first article announcing the measurement of the electronic charge, but was talked out of this by Millikan.

oil ("the highest grade of clock oil"), sprayed into his apparatus with an atomizer. This reduced evaporation from the surface of the droplets, thus keeping their mass constant during experimental runs. More important, Millikan now found that he could observe one individual droplet rather than a cloud, following its motion as it drifted upward or downward many times as the vertical electrical field was turned on or off. For each successive rise and fall of the droplet, the electric charge could be deduced from the speeds of ascent and descent, just as was done by Wilson. (For details, see Appendix G.)

Let us look at one example in detail: drop number 6 in Millikan's 1911 paper.[1] With the electric field turned off the drop fell 0.01021 meters in an average time of 11.88 seconds, so its speed of fall was 0.01021 m/11.88 sec, or 8.59×10^{-4} m/sec. The viscosity of the air was taken by Millikan as 1.825×10^{-5} newton-seconds per meter^2, and the density of the oil was 0.9199×10^3 kilograms per cubic meter. From these data, Millikan calculated that this drop had a radius of 2.76×10^{-6} m and hence a mass equal to 0.9199×10^3 kg/m^3 times $4\pi/3$ times $(2.76 \times 10^{-6}$ m$)^3$, or 8.10×10^{-14} kg. To check Millikan's calculation of the drop radius, note first that the force of gravity is equal to the mass times the common acceleration 9.8 m/sec^2, or

$$8.10 \times 10^{-14} \text{ kg} \times 9.8 \text{ m/sec}^2 = 7.9 \times 10^{-13} \text{ N},$$

and that the viscous drag is given by Stokes's formula as

$$6\pi \times (1.825 \times 10^{-5} \text{ N-sec/m}^2) \times (2.76 \times 10^{-6} \text{ m}) \times (8.59 \times 10^{-4} \text{ m/sec}) = 8.1 \times 10^{-13} \text{ N}.$$

The small discrepancy here arises mostly because Millikan actually used a corrected version of Stokes's law; the correction (discussed in Appendix G) was necessary because air flowing around a very small droplet does not behave strictly like a smooth fluid.

With an electric field of 3.18×10^5 volts per meter, this droplet was seen on its first ascent to rise 0.01021 meters in 80.708 seconds—a speed of 1.26×10^{-4} m/sec. Since it was still the same drop, the drag force was now less than before by just the ratio of velocities

$$\begin{aligned} \text{Drag force} &= \left(\frac{1.26 \times 10^{-4} \text{ m/sec}}{8.59 \times 10^{-4} \text{ m/sec}}\right) \times (8.1 \times 10^{-13} \text{ N}) \\ &= 1.2 \times 10^{-13} \text{ N}, \end{aligned}$$

but since the drop was rising this force now acted downward, in the same direction as gravity. The sum of the gravity and drag forces was then $(7.9 + 1.2) \times 10^{-13}$ newtons, or 9.1×10^{-13} newtons. This had to be just balanced by the upward electric force, which equaled the unknown charge times the electric field of 3.18×10^5 volts per meter. Thus, the charge on the oil drop could be calculated as

$$\frac{9.1 \times 10^{-13}}{3.18 \times 10^5} = 29 \times 10^{-19} \text{ C}.$$

Using raw numbers that had not been rounded off, and taking all corrections into account, Millikan found a more precise value of 29.87×10^{-19} coulombs for the charge of the drop during this ascent.*

Here is the list of the electric charges that Millikan found for this droplet on successive ascents with the electric field on, in units of 10^{-19} coulombs: 29.87, 39.86, 28.25, 29.91, 34.91, 36.59, 28.28, 34.95, 39.97, 26.65, 41.74, 30.00, 33.55. These charges are fairly large multiples of the electronic charge, and it is not so easy to see that they are all whole-number multiples of the same elementary charge. However, the changes in the electric charge from one ascent to the next are much smaller. Taking the difference between each charge and the charge for the previous ascent gives the following *changes* in charge, in units again of 10^{-19} coulombs: 9.91, -11.61, 1.66, 5.00, 1.68, -8.31, 6.67, 5.02, -13.32, 15.09, -11.74, 3.35. It is now fairly obvious that these changes in electric charge are whole-number multiples of a minimum charge, equal to about 1.665×10^{-19} coulombs. In units of this minimum charge, the preceding changes in the charge of the oil drop from one ascent to the next are 5.95, -6.97, 1.00, 3.00, 1.01, -4.99, 4.01, 3.02, 8.00, 9.06, -7.05, and 2.01. The interpretation is that the electron has a charge of about 1.665×10^{-19} coulombs, and that on successive ascents the drop lost six electrons or negative ions, then gained seven, then lost one, then lost three, then lost one, then gained five, and so on.

---

*I have taken some liberties here in presenting Millikan's results in what I thought would be a clearer way. For one thing, Millikan expressed charges in electrostatic units (statcoulombs); I have converted them to coulombs because these are the units used in the rest of this book. Also, Millikan did not actually present a calculation for the electric charge of the oil drop measured each time the drop traveled upward in the electric field. Instead, he presented values of certain quantities that appear in the calculation of the charge and that change from one ascent to another, leaving out common factors that remain constant for a specific drop. I have multiplied by these factors to obtain the actual values for the electric charge that Millikan would have found from his data if he had calculated them. Millikan also included a small correction for the buoyancy of the air, which I have neglected here.

By repeating this experiment for many oil drops, Millikan obtained an average value for the electronic charge of 1.592×10^{-19} coulombs, with an experimental uncertainty of about 0.003×10^{-19} coulombs. This was at the time by far the most accurate measurement of the electronic charge, direct or indirect. Almost more important was the way it was done: in following the oil drop over many ascents and descents, Millikan could watch it gain or lose small numbers of electrons, sometimes only one. The measurements of Townsend, Thomson, and Wilson at the Cavendish Laboratory had really determined only the average ionic charge for the droplets in their clouds of water vapor, and had left open the possibility of a fairly wide range of charges of individual ions or electrons. After Millikan's experiment this was no longer a possibility; every time a drop of oil gained or lost an electric charge, it was always within a percent or so of a whole-number multiple of the same fundamental charge.*

Millikan was quick to use his value for the electric charge to calculate the other atomic parameters. In particular, the Faraday (Avogadro's number times the electronic charge) had been measured in electrolysis to be 96,500 coulombs per mole. Dividing this by the electronic charge, Millikan calculated Avogadro's number to be 96,500 divided by 1.592×10^{-19}, or 6.062×10^{23} molecules per mole. Equivalently and somewhat less abstractly, we could say that electrolysis had given a value for the mass/charge ratio of the hydrogen ion of 1.045×10^{-8} kilograms per coulomb, and the electric charge of the ion was now known to be 1.592×10^{-19} coulomb, so the mass of the hydrogen ion could be determined as the product, 1.663×10^{-27} kilograms. From the known value of about 0.54×10^{-11} kilograms per coulomb for the mass/charge ratio of the electron, the mass of the electron could now be calculated as about 0.54×10^{-11} kg/C times 1.592×10^{-19} C, or 9×10^{-31} kilograms.

It was now easy to estimate the size of atoms. For instance, gold has atomic weight 197 and hydrogen 1.008, so the gold atom has a mass of $197/1.008$ times that of the hydrogen atom, or 3.250×10^{-25} kilograms. The density of gold is 1.93×10^{4} kilograms per cubic meter, so there must be $1.93 \times 10^{4}/3.250 \times 10^{-25} = 5.94 \times 10^{28}$ gold atoms per cubic meter. That is, each gold atom occupies a volume of 1 divided by $5.94 \times 10^{28} = 1.68 \times$

---

* As Holton showed in a study of Millikan's notebooks, Millikan did exercise considerable discretion in choosing which oil drops to include in his published work. Another experimentalist, Felix Ehrenhaft of the University of Vienna, persistently found some drops with anomalously small charges. Time has borne out Millikan's judgment, though Ehrenhaft remained unconvinced until his death.

10^{-29} cubic meters. Taking the cube root, we see that if the gold atoms are tightly packed their diameter must be 2.6×10^{-10} meters.

For many years the Millikan measurement of the electronic charge provided the most accurate basis for the atomic scale. The largest change has resulted from a remeasurement of the viscosity of the air in the 1930s. At present the best value of the electronic charge is $1.6021892 \times 10^{-19}$ coulombs, with an uncertainty of 46 in the last two decimal places. This is less than 1 percent higher than the value obtained in 1913 by Millikan.

Notes

1. R. A. Millikan, "On the Elementary Electrical Charge and the Avogadro Constant." *Physical Review* **32** (1911), 349.

4

The Nucleus

4

The Nucleus

Atoms are electrically neutral, but the electrons discovered by Thomson carry a negative electric charge. If atoms contain electrons, then they must also contain some other material that carries a positive charge to cancel the electrons' negative charge. The great task after the discovery of the electron was to identify this positive material and to describe how it and the electrons are arranged within the atom.

In his 1903 Silliman Lectures at Yale, Thomson suggested that the electrons are stuck in a continuous matrix of positively charged matter, like raisins in a plum pudding. At almost the same time, in Tokyo, Hantaro Nagaoka (1865–1950) was proposing a "Saturnian model," according to which the electrons revolve in orbits around a central positively charged body, like the rings around Saturn or the planets around the sun. We know now that Nagaoka was more nearly right: The positive charge of the atom is indeed concentrated in a small dense nucleus, about which the electrons revolve. But this had to be found by experiment.

The nucleus of the atom was discovered in experiments carried out at the University of Manchester in 1909–11 under the leadership of Ernest Rutherford. Rutherford was born in 1871 at Brightwater in New Zealand to a family of early immigrants from Britain who had settled in a pleasant valley to raise some flax and many children. He was educated in New Zealand at Nelson College, where he became Head Boy, and as an undergraduate at Canterbury College in Christchurch, where he won first-class honors in physics and mathematics. There he began research on electromagnetism, research whose only historical importance is that it earned for him a scholarship of £150 per year, which enabled him in 1895 to come to the Cavendish Laboratory.

During the next few years, while Rutherford was at Cambridge, the world of physics was exhilarated by a rapid sequence of revolutionary developments, culminating in Thomson's 1897 discovery of the electron. First was the discovery of x rays in November 1895 by Wilhelm Konrad Röntgen (1845–

Sir Ernest Rutherford.

Rutherford's first laboratory, in the basement of Canterbury College in New Zealand.

Rutherford in his laboratory at McGill University, Montreal, in 1905.

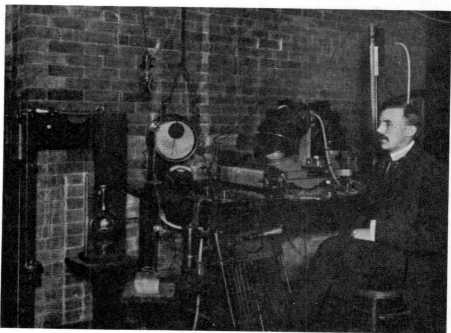

1923) at Würzburg. In brief, Röntgen found that when cathode rays hit the glass wall of a cathode-ray tube, mysterious highly penetrating rays (termed x rays by Röntgen) were emitted, and that these rays could fog photographic plates and cause various materials to fluoresce. We now know that x rays are just light of very short wavelength (typically thousands of times shorter than the wavelengths of visible light) that is emitted when electrons in the outer part of atoms drop into the inner orbits to replace electrons knocked out of the atom by the cathode ray. The discovery of x rays is a little off the main track of our story, but it alerted physicists everywhere to the possibility that there might be various undiscovered forms of radiation.

The next of the stunning discoveries of this period was crucial to Rutherford. Early in 1896 the discovery of radioactivity was announced by Henri Becquerel (1852–1908) in Paris. We will go into the details of this discovery and the early work on the nature of radioactivity in the next section; here it is enough to say that the atoms of radioactive substances emit various kinds of particles with energies millions of times greater than the energies liberated when atoms participate in ordinary chemical reactions.

As might be expected of someone working in Thomson's laboratory, Rutherford was at first interested in the effects of radioactivity and x rays on the conduction of electricity in gases. The energetic particles from radioactive atoms knock electrons out of atoms, and these electrons can then serve as carriers of electric currents. In 1898, after working with Thomson on the effect of x rays on electrical conduction in gases, Rutherford showed that x rays and radioactivity act in essentially the same way. He also recognized at least two kinds of radioactivity, which he called alpha and beta rays.

This work earned Rutherford an appointment as research professor at the newly endowed Macdonald Physics Laboratory at McGill University in Montreal. He sailed for Canada in September 1898, taking care before he left to have some radioactive salts of thorium and uranium sent to him in Montreal. At McGill he formed a partnership with a young chemist from Oxford, Frederick Soddy (1877–1956). During the years at McGill, Rutherford and Soddy worked out the nature of the different kinds of radioactivity, research that will be reviewed in the next section. Rutherford also found time to return to New Zealand in 1900 to get married, to give the Bakerian Lecture at the Royal Society in London in 1903, and to follow Thomson as Silliman Lecturer at Yale in 1905. Although his work had gone more than well at McGill, Rutherford felt isolated from the centers of physics research in Europe, and he seized the chance to return to England when a professorship at the University of Manchester was offered him in 1906. At that time Manchester and Cambridge were the two leading centers of physics research in Britain.

In 1907 Rutherford began a new career at Manchester, where the direction of his work shifted away from the nature of radioactivity itself, and toward its use as a means of settling the question raised at the beginning of this chapter, that of the distribution of matter and charge within the atom. The approach that Rutherford and his colleagues at Manchester took to this question has become an indispensable part of physics: They directed a beam of energetic particles into a thin metal foil, and deduced the distribution of electric charge within atoms of the foil from the probabilities for these particles being scattered by the foil at various angles. (We will analyze these experiments later on in this chapter.) The energetic particles that now are used as probes in such experiments are provided by giant accelerators, such as the ones at Batavia (near Chicago), Geneva, Hamburg, and Stanford—machines whose size is measured in kilometers and which use as much electric power as a good-sized city. The purpose of such experiments has also shifted, away from the study of the structure of the atom and toward the study of the structure of the particles within the atom, or even the particles within these particles. The basic ideas behind the study of scattering as a probe of structure are, however, much the same as those used by Rutherford. Of course, no large accelerators existed in Rutherford's time, and for probes he had to use the energetic particles emitted by naturally radioactive substances. But he settled the question of the arrangement of charge within the atom: The positive charge is concentrated in a small nucleus about which the electrons revolve.

Rutherford's work raised new questions as formidable as the ones he had answered. What determines the sizes and energies of the orbits of the electrons in the atom? Why don't the orbiting electrons steadily radiate electromagnetic waves? And if it is the ordinary electrical attraction between unlike charges that keeps the negative electrons in orbit around the positive nucleus, what keeps the nucleus itself from flying apart? These questions could not be answered within the context of the classical theoretical physics of the time, but a first step toward their solution was taken by the young Danish theorist Niels Hendrik David Bohr (1885–1962), who visited Rutherford at Manchester in 1912 and was then brought back there as Reader in Physics in 1914. Bohr's work led directly to the development in the 1920s of quantum mechanics (which lies outside the scope of this book). Rutherford, unfortunately, was not sympathetic with the developing theory of quantum mechanics, regarding it as too theoretical, too far from the experimental realities with which he had worked. Sir Mark Oliphant recalls that after Bohr gave the Scott Lectures at the Cavendish Laboratory on the uncertainty principle, Rutherford remarked "You know, Bohr, your conclusions seem to me as uncertain as the premises on which they are built." And Sir Nevill Mott quotes a story that, during the exciting period in the 1920s when quantum mechanics was being developed, a colleague asked "How is physics these days, Rutherford?" and Rutherford replied "There is only one thing to say about physics, the theorists are on their hind legs and it is up to us to get them down again." As a theorist I naturally tend to deplore this sort of antitheoretical feeling. But in fact theorists and experimentalists generally get along pretty well with each other, and could hardly get along at all without each other. Rutherford's attitude may have been partly due to the fact that his greatest work was done during a time when so little was known about the nucleus that elaborate mathematical theorizing would have been out of place, and whatever theory was called for, Rutherford was quite capable of supplying himself.

In 1919 Rutherford succeeded Thomson as the Cavendish Professor of Experimental Physics at Cambridge, having applied by telegram on the day of the election. At Cambridge he guided a group of younger men who in the 1930s would open a new era of nuclear physics, especially through the discovery of the neutron by James Chadwick and the disintegration of nuclei by artificially accelerated particles by John D. Cockcroft (1897–1967) and E. T. S. Walton (b. 1903). Rutherford himself was heaped with all the honors that a scientist could earn: the Nobel Prize in Chemistry in 1908 for his work on radioactivity; countless honorary degrees; knighthood in 1914; Presidency of the Royal Society in 1925; and elevation to the peerage in 1930. Recalling his origins, he chose the title of "Baron Rutherford of Nelson" and in the crest of his coat of arms he included a kiwi bird. (The blazoning reads, "On a Wreath of the Colours upon a rock a Kiwi proper.") He remained in active leadership of the Cavendish until his death in 1937.

To my own generation of physicists who did not know Rutherford in person, he leaves behind an impression of acerbity, energy, and frugality. He was no mandarin, but he could be fierce in his judgments. When I visited Cambridge in 1962, I was shown on a wall a carved figure of a crocodile, and was told that it was a symbol for Rutherford.* He was intensely proud and protective of his "boys," the brilliant group of younger experimentalists at the Cavendish, including Blackett, Feather, Kapitza, and Oliphant, as well as

---

* The crocodile is visible in the frontispiece photograph. This figure was carved at Kapitza's request by Eric Gill, who was well known in the 1930s for his sexual preoccupations as well as for his sculptures. I have heard several theories about the association of the crocodile with Rutherford. Gerald Holton tells me of a remark of George Gamow that Rutherford's loud and distinctive voice served as a warning to his students and assistants that Rutherford was coming down the corridor, in the same way that in *Peter Pan* the ticking of the clock swallowed by a crocodile served to warn Captain Hook that the crocodile was after him again. On the other hand, I. Bernard Cohen points out that in the Middle Ages the crocodile was a symbol of alchemy, and Rutherford was fond of

Chadwick, Cockcroft, and Walton. He worked incessantly; if his work at McGill, Manchester, and Cambridge were divided among three different men, they could each be said to have had unusually productive careers in science. He is often associated with the idea that physics should be done with limited resources, with "string and sealing wax," and once when a young physicist complained to Rutherford that he was not getting the apparatus he needed, Rutherford replied "Why, I could do research at the North Pole." However, Rutherford was quite aware of the need for funds, and when he came to the Cavendish in 1919 tried unsuccessfully to raise £200,000 for new facilities. In later years he frequently urged the construction of machines that could accelerate particles to higher and higher energies.

At the symposium on *Nuclear Physics in Retrospect,* Maurice Goldhaber commented on the increase in the scale of experiments in nuclear physics: "The first to disintegrate a nucleus was Rutherford, and there is a picture of him holding the apparatus in his lap. I then always remember the later picture when one of the famous cyclotrons was built at Berkeley, and all of the people were sitting in the lap of the cyclotron." The scale of modern elementary particle physics is even larger. The accelerator at Fermilab is a ring four miles in circumference, surrounding a fair-sized bit of Illinois prairie on which a herd of bison peacefully graze. It is sometimes asked why physicists today need hundreds of millions of dollars for giant accelerators when Rutherford accomplished so much on a table top. I think the answer is that those discoveries on the fundamental nature of matter that can be made with string and sealing wax have mostly been made, in large part by Rutherford.

So far, I have emphasized the problem of the distribution of electric charge in the atom, but the work of the Rutherford group at Manchester settled another question that had been raised by Thomson's discovery of the electron: What is the distribution of mass in the atom? As we saw in Chapter 3, the work of John Dalton and other chemists in the early nineteenth century had determined the *relative* masses of the atoms of different elements, revealing that the carbon atom weighs 12 times as much as the hydrogen atom, the oxygen atom weighs 16 times as much as hydrogen, and so on. Also, the work

---

comparing himself with alchemists; one of his books is called *The Newer Alchemy,* and in his office at the Cavendish there hung a woodcut of an alchemical laboratory, with a stuffed crocodile hanging over the alchemical apparatus. The only mention of the Rutherford crocodile that I can find in print is in the book by A. S. Eve, who suggests that it may symbolize Rutherford's acumen and career, because the crocodile never turns back. Brian Pippard adds one more suggestion, that in Russian, Kapitza's native language, *krokodil* is slang for "boss." I had the opportunity to ask Kapitza about the significance of the crocodile, when we were together at a meeting by Lake Constance. He said it is a secret.

on electrolysis by Faraday and others had shown that the electrically charged atoms (ions) that carry electric currents in solutions of acids or salts have a mass/charge ratio of about 10^{-8} kilograms/coulomb for hydrogen, and proportionally more for the ions of heavier atoms. After the discovery of the electron, it became fairly clear that these ions are nothing but atoms that have gained one or more electrons (for negatively charged ions) or lost one or more electrons (for positively charged ions). On this basis, the electric charge of the hydrogen ion should be just equal in magnitude to the charge of the electron. Thus, since the mass/charge ratio of the electron is 1/2,000 that of the hydrogen ions, and the charges are equal, the mass of the hydrogen ion (and atom) should be some 2,000 times larger than the mass of the electron. Does this mean that atoms consist of thousands of electrons? Or is most of the mass of the atom somewhere else, perhaps associated with the positive electric charges?

As we will see later, the 1909–11 Manchester experiments showed not only that the positive electric charge of the atom is concentrated in a tiny nucleus, but also that this nucleus contains almost all the mass of the atom. What then does the nucleus consist of? Dalton found that atoms have masses generally close to multiples of the mass of the hydrogen atom, so one might think that the nuclei of atoms consist of heavy, positively charged particles that can be identified with the nucleus of hydrogen, particles that Rutherford in 1920 named *protons*. However, Rutherford's own results showed that this would not work; for instance, the mass of the helium nucleus is four times that of hydrogen, but Rutherford found that its electric charge was only twice as great. As we shall see, Moseley measured other nuclear charges in 1913 and found the same pattern—for instance, calcium has an atomic weight 40 times that of hydrogen, but a nuclear charge only 20 times as great. Throughout the teens and the twenties, most physicists thought that the nucleus also contained electrons; for instance, the helium nucleus would consist of four protons (explaining the mass) plus two electrons (to cancel two units of charge). This was wrong, and the correct answer was not found until the discovery in 1932 of the neutron, the last of the subatomic particles.

The Discovery and Explanation of Radioactivity

The proportion of scientific discoveries achieved by accident is not as large as many people think. However, there is no question about the accidental nature of one of the great discoveries that opened up the physics of the twentieth century: the discovery of radioactivity.

In February 1896, Antoine Henri Becquerel (1852–1908), professor of physics at the Ecole Polytechnique, was exploring the possibility that sunlight might cause crystals to emit penetrating rays like the x rays that had been discovered by Röntgen a few months earlier. Becquerel's method was simple: Various crystals were placed next to photographic plates wrapped in dark paper, with a copper screen separating them. If sunlight caused the crystals to emit rays similar to x rays, then these rays might penetrate the dark paper around the plates but not the copper wires of the screen, so that when the plates were developed they would be found to be exposed, except for an unexposed silhouette of the copper screen.

As luck would have it, one of the crystals studied by Becquerel was composed of a salt of uranium, uranium-potassium bisulfate. (Becquerel had guessed that the effect he was looking for might be associated with phosphorescence, and these uranium salts were known to be phosphorescent.) Also by chance, at just that time the weather was none too good. Here is Becquerel's own account (reported later that year) of what happened:

> [On 26 and 27 February] the sun appeared only intermittently [so] I stopped all experiments and left them in readiness by placing the unwrapped plates in the drawer of a cabinet, leaving in place the uranium salts. The Sun did not appear on the following days, and I developed the plates on March 3, expecting to find only faint images. The silhouettes appeared, on the contrary, with great intensity. . . .

Two months later, Becquerel noted the following:

> From March 3 to May 3 the salts were kept in a lead-walled box, which was kept in the darkness. . . . Under these circumstances, the salts continued to emit active radiation. . . . All the uranium salts I have studied, whether phosphorescent or not under light or in solution, gave me corresponding results. I have thus been led to the conclusion that the effect is due to the presence of the element uranium in these salts.

Becquerel was right in attributing these rays to uranium, and for a few years they were known in France as *rayons uranique*. However, other elements could produce such rays. In 1898 Marie Sklodowska Curie (1867–1934) in Paris discovered that similar rays were given off by the element thorium, and she and her husband Pierre Curie (1859–1906) discovered the element radium, finding it to be millions of times more active than uranium. That year the Curies gave the phenomenon its modern name, *radioactivity*.

But what was radioactivity? One great complication in dealing with this problem was that three different kinds of ray are emitted by radioactive atoms.

Marie and Pierre Curie in their laboratory.

As mentioned above, Rutherford's research on radioactivity at the Cavendish Laboratory between 1895 and 1898 showed that there are at least two different kinds of radiation, which he called alpha and beta rays. The beta rays were about as penetrating as x rays, but the alpha rays had much less penetrating power; most of them were stopped by an aluminum foil 0.001 inch thick. Becquerel (and, independently, F. Giesel) noted in 1899 that part of the radiation emitted by uranium (the type called beta radiation by Rutherford) could be deflected by a magnetic field, in the same direction as cathode rays. Using a method like Thomson's, Becquerel measured the beta rays' mass/charge ratio and found it to be close to the value measured for electrons by Thomson. (The measurement was made with greater precision in 1907 by Kaufmann.) It was clear that beta rays are just electrons, but electrons with speeds much higher than those in cathode rays.

Alpha rays were more difficult to deflect with electric or magnetic fields, but in 1903 Rutherford (then at McGill) succeeded in measuring this deflection and used it to determine the mass/charge ratio of the alpha particles as roughly equal to that for the hydrogen ion in electrolysis. When these experiments were repeated with greater precision in 1906, Rutherford found that the mass/charge ratio for alpha particles is actually about twice that for hydrogen ions. This might mean that the alpha particles were ions with an electric charge equal to that of the hydrogen ion and an atomic weight of 2 (twice that of hydrogen). However, no chemical element was known with an atomic weight of 2. Rutherford soon guessed that alpha particles are instead ions of helium, the next lightest element after hydrogen, which has an atomic weight of 4; then, since the mass/charge ratio is twice that for hydrogen ions and the mass is four times larger, the charge must be twice that of a hydrogen ion— equal in magnitude but opposite in sign to *two* electron charges. Rutherford had made the first determination of what later became known as an atomic number; the helium ion emitted in alpha radioactivity has an electric charge of +2 in units of the hydrogen-ion charge, because this is the charge of the helium nucleus, and the alpha particles emitted by radioactive substances are just helium nuclei, missing the two electrons that are its normal complement.

The identification of alpha rays with helium ions was suggested to Rutherford at least in part by the fact that helium was known to be associated with radioactive materials. In fact, the first discovery of helium on earth was made in 1895 by the British chemist William Ramsay (1852–1916), who found it in the uranium mineral cleveite. I say "on earth" because helium was actually found first on the sun. When a narrow beam of light from the sun is passed through a prism and observed with a telescope, the resulting spectrum is seen to be crossed by a number of bright and dark lines which are produced

Dark absorption lines in the part of the solar spectrum between 3900 angstroms and 4600 angstroms in wavelength.

when light of certain definite wavelengths is emitted or absorbed by atoms on the sun's surface. Most of these lines could be identified with similar lines produced by various elements in terrestrial laboratories, but one spectral line that was first observed in the eclipse of 1868 remained mysterious. The astronomer J. Norman Lockyer (1836–1920) decided that it was due to a new element, which was then named *helium* after the Greek word "helios" for the sun. Helium is in fact very common on the sun and in the universe in general, making up about one-fourth of the mass of most stars. It is rare on the earth only because the helium atom is so light and so inactive chemically. Individual helium atoms in our atmosphere can easily be given high enough speeds by collision with air molecules to escape the earth's gravity, and helium cannot be trapped into relatively heavy molecules in the way that hydrogen is trapped in water.

The conclusion that helium is produced in radioactivity became inescapable when Ramsay and Soddy, at McGill in 1903, observed helium being formed by radium salts. Finally, in 1907–1908, Rutherford, together with T. D. Royds at Manchester, was able to collect enough of the alpha particles emitted by a sample of radium to observe the same spectral lines by which helium is identified on the sun, thus confirming indisputably that alpha particles are helium ions. Rutherford did not know it then, but the reason that alpha particles are commonly emitted by radioactive atoms is the same as the reason that helium is common in our universe: The helium nucleus is by far the most tightly bound of the lightest atomic nuclei.

The third of the three kinds of radioactivity were rays that (like beta rays or x rays) were highly penetrating, but (like alpha rays or x rays) could not easily be deflected by a magnetic field. First observed in France by P. Villard in 1900, they were named *gamma rays* by Rutherford in 1903. Rutherford guessed that gamma rays are, like x rays, light of very short wavelength, but this was not proved until 1914, when Rutherford, together with E. N. da Costa Andrade (1887–1971), succeeded in measuring the wavelength of gamma rays by observing their scattering from crystals. (Gamma rays were much less important in the early history of radioactivity than alpha or beta rays, and I will not have much to say about them here.)

So alpha rays are doubly charged helium ions (actually, helium nuclei), beta rays are electrons, and gamma rays are pulses of light. But what causes atoms to emit such rays? An important clue was discovered by Rutherford in 1899, soon after his arrival at McGill. He had observed a year earlier that the radioactivity from thorium sometimes seemed to fluctuate, especially if the thorium was in a draft of air. By blowing air over the surface of a sample of thorium and into a flask, he was able to collect a gas, which he called "thorium emanation." (A similar gas emitted by radium had been observed a little earlier by Friedrich Ernst Dorn.) The gas was intensely radioactive, and evidently accounted for part of the radioactivity that had been attributed to thorium itself. (Incidentally, in all these experiments at McGill, the amount of radioactivity was measured through its effect on the conduction of electricity by gases—the same phenomenon on which Rutherford and Thomson had worked together at Cambridge.)

Part of the importance of this discovery lay in what it revealed about the complexity of the phenomenon of radioactivity. Much of the radioactivity associated with elements such as thorium or uranium is due to tiny amounts of substances such as thorium emanation or radium emanation which are themselves produced by the radioactivity of the parent element (or by the radioactivity of other substances that are in turn produced by the radioactivity of the parent element). For instance, Rutherford and Soddy found in 1903 that 54 percent of the radioactivity of thorium (and all the production of thorium emanation) is due to a highly radioactive substance that they called "thorium X." The thorium X could be concentrated in a solution of a thorium salt (thorium nitrate) by adding ammonia to the solution; the thorium would separate out as a precipitate (thorium hydroxide), leaving the thorium X behind in the liquid. With thorium X separated in this way, the precipitate of thorium was much less radioactive and no longer produced thorium emanation. However, during the Christmas vacation of 1901 a sample of thorium from which the thorium X had been removed was allowed to stand for three weeks, and on returning to the laboratory Rutherford and Soddy found that the thorium X had built back up to its normal abundance. The sample had recovered both its radioactivity and its production of thorium emanation. The conclusion was that thorium X is not merely an impurity that happens to accompany natural thorium, but is actually produced by the thorium, just as the thorium emanation is produced by the thorium X.

Even more important than this untangling of the complexities of radioactivity was the realization that these various substances produced by radioactivity were actually elements different from the original radioactive element. In 1902 Rutherford and Soddy showed that thorium emanation is a new "noble gas," a member of the family of chemically inert elements discovered shortly before by Ramsay (the family also includes helium, neon, argon, krypton, and xenon). The new element was at first called "niton," but later came to be known as *radon*. Radium emanation was also found to be a form of radon. (In modern terms, thorium emanation and radium emanation are two different isotopes of radon, ^{220}Rn and ^{222}Rn. There are altogether some twenty known isotopes of radon.) Also, thorium X was clearly a different chemical element from thorium; it was later identified as an exceptionally radioactive form of the element radium. (Thorium X is ^{224}Ra, while ordinary radium is ^{226}Ra.) So thorium turned into a form of radium which then turned into a form of radon.

In a pair of classic 1903 papers[1] entitled "The Cause and Nature of Radioactivity," Rutherford and Soddy explained that radioactivity is actually a change of one chemical element into another, a change caused by the emission of a charged alpha or beta particle. This was strong stuff—the immutability of the elements had become an axiom of chemistry. The next year, Rutherford described the Rutherford-Soddy "disintegration theory" before a skeptical audience at the Royal Society in London. In the audience was Pierre Curie, then himself preparing a survey of research on radioactivity. In his survey, Curie did not even mention the disintegration theory of Rutherford and Soddy.

Thorium emanation provided one other crucial insight into the nature of radioactivity. Rutherford noticed that the intensity of the radioactivity it produced decreased rapidly: After about one minute a given sample of gas has only half its original radioactivity, after two minutes only one-fourth, after three minutes only one-eighth, and so on. His interpretation, as explained in the papers with Soddy, was that each atom of thorium emanation has a 50 percent probability of emitting an alpha particle in each minute (actually, in each 54.5 seconds), regardless of how long the atom has already survived or how many other atoms are present, and when it does emit an alpha particle it ceases to be an atom of thorium emanation. (Of course, the alpha-particle emission does not take place only at intervals of 54.5 seconds; it can occur at any time.) If one starts with a definite amount of thorium emanation, then after 54.5 seconds half of it is gone, so the sample's radioactivity has half its original intensity. After another 54.5 seconds, half the remaining thorium emanation has disappeared, so the radioactivity has half of one-half, or one-fourth, of its original intensity, and so on. The important points were that because the rate at which atoms of thorium emanation emit alpha particles does not depend on the presence of other atoms this must be a one-atom process, unlike an ordinary chemical reaction, and that because its rate does not depend on the previous history of the atom alpha-particle emission must be a probabilistic process, like flipping a coin. It is an old fallacy that if a coin is flipped many times and always comes up heads, then on the next flip it is somehow more likely to come up tails. This is not so—if the coin is perfectly symmetrical the chance of its coming up heads is 50 percent on each toss, so the chance of its coming up heads on two tosses is 25 percent, on three tosses 12.5 percent, and so on. In the radioactive decay of an atom of thorium emanation it was as if a coin were being flipped every 54.5 seconds, with the atom surviving only as long as the coin came up heads. (But radioactive decay is unlike flipping a coin in that it can occur at any time.) The reason for this probabilistic behavior was not understood fully until quantum mechanics was applied to nuclear physics in the late 1920s and the early 1930s.

Other radioactive elements were soon found to follow a similar decay law. For each one, there is a characteristic *half-life,* which is the time in which an atom has a 50 percent chance of suffering radioactive transformation, or equivalently the time in which the radioactivity of a sample of the element will lose half its intensity.* For thorium emanation, as we have seen, the half-life is 54.5 seconds; for radium emanation it is 3.823 days; for thorium X it is 3.64 days; and so on. (These half-lives provided one of the hints that led to the discovery of isotopes, discussed in Chapter 3. Thorium emanation and radium emanation are the same element, radon, but have very different half-lives.) The reason that radioactivity had not been observed to decrease in thorium, uranium, or radium is that these elements (or, more precisely, their most common isotopes) are extremely long-lived: The half-life of radium (^{226}Ra) is 1,600 years, that of thorium (^{232}Th) is 1.41×10^{10} years, and that of uranium (^{238}U) is 4.51×10^9 years. It is true that the radioactivity observed in samples of radium, thorium, or uranium is largely due to small amounts of highly active elements like thorium X, which are very short-lived, but these rapidly decaying elements are continually replenished by the radioactivity of their parents, so the half-life observed in an undisturbed sample of thorium or radium or uranium is the long half-life of the parent element. When Rutherford and

---

* There is nothing special about the number one-half, or 50 percent. One could just as well speak of a "third-life," the time in which the radioactivity of a given sample of this element would be reduced to ⅓ of its original intensity, or equivalently the time in which an individual atom has a 66⅔ percent probability of undergoing a radioactive disintegration. Since ⅓ = (½)$^{1.58}$, the third-life is 1.58 half-lives. In fact, it is very common to describe radioactive disintegration not in terms of half-lives (or third-lives), but in terms of mean lives, the average length of time that each atom survives before undergoing a radioactive disintegration. It is shown in Appendix H that the probability of the atom undergoing a disintegration in a short time interval equals the ratio of the time interval to the mean life. Also, the mean life is 1.443 half-lives. For instance, radium has a half-life of 1,600 years, so its mean life is 1.443 × 1,600 years, or 2,310 years. In one year the probability that any given radium atom will decay is then 1 year/2,310 years, or 0.04 percent.

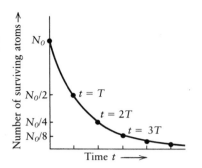

Half-life decay curve. After each time interval T, half of the remaining atoms have decayed.

Soddy removed the thorium X from a sample of ordinary thorium, the radioactivity was at first greatly reduced, but in a few days it increased as the decay of thorium replenished the thorium X, until the amount of thorium X became large enough that just as much was decaying each second by emitting alpha particles as was being produced by the decay of thorium. After this, the amount of thorium X decreased imperceptibly with the 1.41×10^{10}-year half-life of its parent. At the same time, the thorium X originally removed from the sample of thorium would be losing its radioactivity with its characteristic half-life of 3.64 days. When elevated to the peerage in 1930, Rutherford included these rising and falling curves of radioactive intensity in his baronial coat of arms, beneath the kiwi.

The reader may have wondered how some radioactive elements like radium, which are found in the earth's crust, can have half-lives (1,600 years for radium) that are much shorter than the age of the earth? The answer is that all such elements are produced by the radioactive decay of longer-lived elements; in the case of radium, the parent is uranium. The only radioactive elements not produced in this way are those with half-lives of at least several hundred million years: uranium and thorium. Even for these long-lived elements, the abundances we observe strongly reflect the rate of radioactive decay. For instance, uranium has two long-lived isotopes, ^{238}U and ^{235}U, with half-lives of 4.51×10^9 years and 7.1×10^8 years, respectively. It is believed that these isotopes were produced in roughly equal amounts in the explosions of an earlier generation of stars, which injected uranium into the interstellar medium for which the solar system was formed. Today on earth we observe that ^{235}U is only about 0.0072 times as abundant as ^{238}U. The conclusion is that the uranium was formed so long ago that most of the shorter-lived ^{235}U has decayed. To be more quantitative, we note that 0.0072 is roughly equal to $(\frac{1}{2})^7$—that is, $\frac{1}{2}$ times itself seven times—so the difference in the number of ^{235}U and ^{238}U half-lives that have passed since the uranium was formed must be about 7. This yields a uranium age of about 6×10^9 years, because then the

^{235}U is about 8.5 ^{235}U half-lives old and the ^{238}U is about 1.5 ^{238}U half-lives old, and the difference is seven half-lives. (Formulas for doing this sort of calculation are worked out in appendix H.) This little calculation provides our most reliable means of setting a lower bound on the age of our universe: It must be at least about 6×10^9 years old.

How is it possible to determine half-lives as long as those of natural uranium and thorium, which are measured in billions of years? The answer is definitely not that one waits a while and sees the radioactivity decrease—the decrease is much too slow. For instance, the nine years that Rutherford was at McGill represent a fraction of the thorium half-life of

$$\frac{9 \text{ years}}{1.41 \times 10^{10} \text{ years}} = 6.4 \times 10^{-10},$$

so the radioactivity of the sample of thorium that Rutherford had sent to him when he sailed for Montreal had in these nine years been reduced only by the factor

$$\left(\frac{1}{2}\right)^{6.4 \times 10^{-10}} = 0.99999999956.$$

This decrease could not be observed even with the best techniques available now. Instead, one must measure the half-life by counting the radioactive disintegrations of individual atoms, for example by counting the scintillations produced when alpha particles from the decaying atoms strike a zinc sulfide screen. Dividing the number of disintegrations per second from a given sample of a radioactive element by the number of atoms in the sample (determined by multiplying Avogadro's number by the number of grams and dividing by the atomic weight), one obtains the probability that an individual atom will undergo a radioactive disintegration in one second. The half-life is then calculated as the time required for these probabilities to build up to 50 percent. In this way half-lives have been measured that are vastly longer than the age of the Earth; the longest measured so far is that of technetium 122, which is about 10^{22} years. At present a number of experimental groups are looking for a possible faint radioactivity of elements like hydrogen or oxygen (which have generally been believed to be absolutely nonradioactive) by observing up to 5,000 tons of ordinary materials such as iron or water, waiting for the sudden appearance of the charged particles that would be produced by such radioactive decays. Since 5,000 tons of water contain 1.5×10^{32} water molecules

(4.5×10^9 grams times Avogadro's number of 6×10^{23} divided by the molecular weight of 18), a decay probability of 10^{-31} per molecule per year would yield 15 decay events per year, which ought to be detectable. This would correspond to a half-life of about 10^{32} years for individual nuclear particles.

Incidentally, there are some radioactive elements (such as radium) whose half-lives are short enough to be measured from the rate at which their radioactivity decreases, and yet long enough also to be measured by assembling a sample of known mass and counting the radioactive disintegrations. The half-lives measured in these two ways must of course agree if one has correctly calculated the number of radioactive atoms in the sample. Alternatively, one can use the half-life measured from the decay of radioactivity together with the number of disintegrations per second per gram of radioactive material to calculate the number of atoms per gram, which immediately (multiplying by the atomic weight) yields Avogadro's number. In this way, by 1909 Avogadro's number had been determined to be about 7×10^{23} molecules per mole, but this result was immediately supplanted by the far more accurate value obtained by Millikan.

I have not yet mentioned the feature of radioactivity that was most disturbing to physicists in the first decade of the twentieth century. In his 1903 experiments on the electrical and magnetic deflection of alpha particles, Rutherford had found that the velocity of alpha particles from radium was about 2.5×10^7 m/sec, or roughly $\frac{1}{10}$ the speed of light. Now, the kinetic energy of any particle is $\frac{1}{2}$ the mass times the square of the velocity, so the kinetic energy per mass of a particle with this velocity is

$$\frac{\text{Kinetic energy}}{\text{Mass}} = \frac{1}{2} \times (2.5 \times 10^7)^2$$
$$= 3 \times 10^{14} \text{ joules/kilogram.}$$

The atomic weight of alpha particles is 4 (although until 1906 Rutherford thought it was roughly 1) and the atomic weight of radium is 226, so each alpha particle has 4/226 the mass of the atom that emits it. The energy released per kilogram of radium when all its atoms have been transmuted into another element by alpha-particle emission is therefore about*

---

*Rutherford actually did this calculation in a more roundabout way. He used the (then poorly known) value of Avogadro's number to estimate the mass of alpha particles, used this mass to calculate the kinetic energy of individual alpha particles (and not just their ratio of kinetic energy to mass), and then divided by the mass of the radium atom (also determined from Avogadro's number) to get the energy produced per mass of radium. It is easy to see that the answer is the same as we calculated here, and is actually independent of the value adopted for Avogadro's number.

$$\frac{4}{226} \times 3 \times 10^{14} = 5 \times 10^{12} \text{ J/kg}.$$

In contrast, the energy released by burning common fuels like natural gas is of the order of 5×10^7 joules per kilogram. The energy released in the radioactive decay of a given mass of radium is thus about 10^5 times that released in ordinary chemical processes. (In 1903 Curie and Laborde measured the heat generated by radioactivity directly; they found that radium, together with its decay products, produces 100 calories per gram per hour—enough to melt itself in a few hours if the heat is not allowed to dissipate.) In a 1904 paper, Rutherford and Soddy concluded that "all these considerations point to the conclusion that the energy latent in the atom must be enormous compared to that rendered free in ordinary chemical change." They then went on to a remarkable speculation that similar enormous energies were stored even in ordinary nonradioactive atoms. In their words, "Now the radio elements differ in no way from the other elements in their chemical and physical behavior. . . . Hence there is no reason to assume that this enormous store of energy is possessed by the radio elements alone." They further proposed that this would solve the ancient puzzle of the source of the energy radiated by the stars: "The maintenance of solar energy . . . no longer presents any fundamental difficulty if the internal energy of the component elements is considered to be available, i.e., if processes of subatomic change are going on."[2]

Rutherford never expressed any doubt that radioactivity obeys the principle of conservation of energy. He recognized that the energy released in the radioactivity of atoms of thorium emanation is just what was stored in these atoms when they were formed by the radioactive decay of atoms of thorium X, and this stored energy together with the energy released in the radioactivity of thorium X is just what was stored in the atoms of thorium X when they were formed by the radioactive decay of the parent thorium atoms. (This was not obvious; among those who speculated that radioactive substances might be drawing their energy from some external source, Pais* lists the Curies, Lord Kelvin, and Jean Perrin.) But in what way do these atoms store such enormous amounts of energy? How did it get stored in the parent atoms of natural thorium? Why is this energy released in a sequence of changes in the chemical element of the atom, each change accompanied by the emission of an alpha or beta particle? These questions could not be answered until after the discoveries of the nucleus and its constituents.

---

* See the Notes for Further Reading at the back of this book.

Rutherford and Hans Wilhelm Geiger.

The Discovery of the Nucleus

Shortly after Rutherford came to Manchester in 1907 he was joined in his work by a young German postdoctoral researcher, Hans Wilhelm Geiger (1882–1945), and an even younger student from New Zealand, Ernest Marsden. Geiger began a program of research on the scattering of alpha particles as they passed through thin metal foils, a phenomenon first studied by Rutherford at McGill in 1906. Alpha particles from a radium source were allowed to fall on a narrow slit in a screen, so that only a narrow beam of alpha particles would emerge from the slit. This beam was then passed through a metal foil, which caused the beam to spread because of the slight bending of the path of the alpha particles as they passed close to the atoms of the foil. The spread in the beam was then measured by allowing the beam to fall on a sheet of zinc sulfide, which emits a visible flash of light when struck by even one alpha particle. Geiger reported in 1908 that the number of scattered particles decreased rapidly with increasing scattering angle, and that no alpha particles were observed to be scattered by more than a few degrees.[3]

So far, there were no surprises. Then in 1909 Rutherford, for some reason, had the idea of checking whether some alpha particles might be scattered by much larger angles from the original direction of the beam. Here is his reminiscence of what happened, as quoted from one of Rutherford's last lectures:

> One day Geiger came to me and said, "Don't you think that young Marsden, whom I am training in radioactive methods, ought to begin a small research?" Now I had thought that too, so I said, "Why not let him see if any alpha particles can be scattered through a large angle?" I may tell you in confidence that I did not believe that there would be, since we knew that the alpha particle was a very fast massive particle, with a great deal of energy, and you could show that if the scattering was due to the accumulated effect of a number of small scatterings the chance of an alpha particle being scattered backwards was very small. Then I remember two or three days later Geiger coming to me in great excitement and saying, "We have been able to get some of the alpha particles coming backwards. . . ." It was quite the most incredible event that has ever happened to me in my life. It was almost as incredible as if you fired a 15-inch shell at a piece of tissue paper and it came back and hit you.[4]

Whether or not Rutherford was really this surprised, many physicists would have been. The reasons will go a long way toward explaining how these large-angle scatterings led in 1911 to Rutherford's conception of the atomic nucleus.

First, as mentioned by Rutherford in the above quote, it was quite impossible to explain the large-angle scattering of an alpha particle in terms of a large number of small-angle scatterings. Geiger and Marsden found in 1909 that the most probable angle by which alpha particles from radium C (the granddaughter of natural radium) are scattered in passing through a thin gold foil (4×10^{-5} cm thick) is 0.87°, but about one in every 20,000 alpha particles is scattered backward (that is, by more than 90°, which is more than 100 times the most probable angle). A well-known theorem in the mathematical theory of probability, the central limit theorem, gives a formula for the probability of finding any specific value for a quantity that is made up of many statistically independent small increments, each of which can be in any direction. According to this formula, the probability of finding such a quantity to have a value

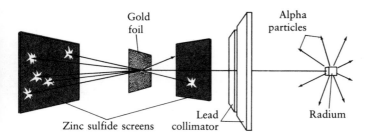

Geiger's and Marsden's scattering experiment with alpha particles and gold foil.

more than 100 times its most probable value (or, strictly speaking, more than 100 times its root-mean-square value) is only 3×10^{-2174}. Even if all the material of the universe consisted of alpha particles and each alpha particle was fired billions of times per second through this gold foil, the chance that such an improbable event would have occurred even once in the history of the universe would still be utterly negligible. As Rutherford concluded, these large-angle scatterings could only be explained if there is an appreciable probability that an alpha particle is deflected by a large angle in a *single* encounter with an atom.

However, the alpha particles carry so much energy that, in order for them to be deflected by a large angle in a single encounter with a charged atomic particle, they must experience enormous electric fields and hence must come extremely close to the charged particle they encounter. We can make this quantitative by a little calculation that Rutherford used for this purpose in his 1911 paper. Consider the especially simple case where an alpha particle is fired directly at some sort of a heavy positively charged particle inside the gold atom, is brought momentarily to reset by the repulsive electric force between it and the atomic particle, and then rebounds, coming straight back toward its source like a rubber ball thrown against a concrete wall. When the alpha particle is far from the positively charged atomic particle, its energy is just its energy of motion (see p. 56):

Initial kinetic energy = $\frac{1}{2}$ × Mass of alpha particle × (Initial velocity of alpha particle)².

At the moment when the alpha particle is brought to rest near the atomic particle, all its kinetic energy has been used up by doing work against the repulsive electric force, so the initial kinetic energy must be equal to the amount of this work. Now, work is force times distance, and Coulomb's law gives the force here as

$$\text{Force} = \frac{k_e \times \text{Charge of alpha particle} \times \text{Charge of atomic particle}}{(\text{Distance between alpha particle and atomic particle})^2},$$

with k_e the universal constant 8.987×10^9 newton-meters^2/coulomb^2 (see p. 35). However the force changes as the distance between the alpha particle and the atomic particle decreases, and the total distance traveled as the alpha particle comes in from infinity is really infinite, so here we cannot simply multiply distance by force to get work. Nevertheless, as Appendix I shows, the work done in bringing the alpha particle in to a given distance from the atomic particle is correctly calculated by just multiplying the above formula by this distance (which cancels one factor of this distance in the denominator):

$$\begin{array}{l}\text{Work done in bringing} \\ \text{alpha particle to a} \\ \text{given distance from} \\ \text{the atomic particle}\end{array} = \frac{k_e \times \begin{array}{c}\text{Charge of} \\ \text{alpha particle}\end{array} \times \begin{array}{c}\text{Charge of} \\ \text{atomic particle}\end{array}}{\text{Distance}}.$$

Now, equating the initial kinetic energy of the alpha particle to the work done in bringing it to its point of closest approach to the atomic particle, we have the equation

$$\tfrac{1}{2} \times \begin{array}{c}\text{Mass of} \\ \text{alpha} \\ \text{particle}\end{array} \times \left(\begin{array}{c}\text{Initial velocity} \\ \text{of alpha particle}\end{array}\right)^2 = \frac{k_e \times \begin{array}{c}\text{Charge of} \\ \text{alpha particle}\end{array} \times \begin{array}{c}\text{Charge of} \\ \text{atomic particle}\end{array}}{\begin{array}{c}\text{Distance between alpha particle and atomic} \\ \text{particle at point of closest approach}\end{array}}$$

It is easy then to solve for the distance at the point of closest approach:

$$\begin{array}{c}\text{Distance between alpha particle} \\ \text{and atomic particle at point of} \\ \text{closest approach}\end{array} = \frac{2 \times k_e \times \begin{array}{c}\text{Charge of} \\ \text{atomic particle}\end{array}}{\begin{array}{c}\text{Mass/charge} \\ \text{ratio of alpha} \\ \text{particle}\end{array} \times \left(\begin{array}{c}\text{Initial velocity of} \\ \text{alpha particle}\end{array}\right)^2}$$

Now we can put in the numbers. In the Geiger-Marsden experiments the velocity of the alpha particles was 2.09×10^7 meters per second, and the
mass/charge ratio of the alpha particles was known to be about 2×10^{-8}
kilograms per coulomb. (Both quantities had been measured by the Thomson
technique of deflection by known electric and magnetic fields.) The charge of
the hypothetical atomic particle was, of course, not known to Rutherford, so
let's suppose that it is a multiple Z of the fundamental electronic charge
1.64×10^{-19} coulombs measured by Millikan. Then the distance between the
alpha particle and the atomic particle at the point of closest approach is

$$\frac{2 \times (8.987 \times 10^9 \text{ N-m}^2/\text{C}^2) \times Z \times (1.64 \times 10^{-19} \text{ C})}{(2 \times 10^{-8} \text{ kg/C}) \times (2.09 \times 10^7 \text{ m/sec})^2} = 3 \times Z \times 10^{-16} \text{ m}.$$

Even if the atomic particle carries an electric charge several hundred times that
of the electron, the distance of closest approach must be less than 10^{-13} meters. This is a very small distance indeed, about 1/1,000 the size of gold atoms
estimated from the density of gold as described in Chapter 3. Evidently the
large-angle scattering of alpha particles is due to encounters not with objects
about as large as an atom, but with very much smaller particles within the
atom.

I have described the head-on encounter of an alpha particle with a
hypothetical positively charged atomic particle, but an alpha particle could
also be deflected straight backward in an encounter with a negatively charged
particle. If we suppose that the alpha particle is fired in such a direction that it
just misses the negative atomic particle, then under the influence of the attractive electric force it will loop around the atomic particle on a narrow hyperbolic orbit and return to infinity along almost the same direction as it came
from, just like a comet that is not bound in the solar system encountering the
sun. In this case the alpha particle would come even closer to a negative atomic
particle that deflects it than it would to a positive particle.

Even though a negatively charged atomic particle could in principle
produce large-angle deflections of alpha particles, Rutherford could be quite
sure that the observed large-angle deflections were not due to encounters with
electrons. Electrons are just too light for encounters with them to have much
effect on the motion of an alpha particle. A billiard ball can experience a large
deflection if it collides with another billiard ball, but not much can happen to
its motion if it rolls into a stationary ping-pong ball, as long as the ping-pong
ball is not glued to the billiard table.

This can be made more quantitative by using one of the great conservation principles of physics, the law of conservation of momentum. The momentum of any particle is defined as the product of its mass and its velocity, so the rate of change of a particle's momentum is equal to its mass times its acceleration (the rate of change of its velocity)—which, according to Newton's Second Law, is just the force acting on the particle. Momentum, like force or velocity or acceleration (but unlike energy or charge), is a directed quantity, and it can be specified by giving its components along three directions (say north, east, and up). Now, Newton's Third Law of Motion says that the force one particle exerts on another is equal in magnitude and opposite in direction to the force that the second particle exerts on the first, so the same must be true of the rate of change of momentum. Hence, the increase in any component of either particle's momentum in any short interval of time is balanced by the decrease in that component of momentum carried by the other particle, and the *total* value of each component of momentum remains constant.

How does this apply to the simple case of an alpha particle that is fired straight at some charged atomic particle at rest and either bounces straight back or continues in the same direction? Here we need to consider the component of momentum in only one direction—the direction of the alpha particle's original motion—so there are two conditions that must be met in the collision: the conservation of this component of momentum and the conservation of energy. With a given initial alpha-particle velocity, there are also two unknowns: the final velocity of the alpha particle and the recoil velocity of the struck atomic particle. With two conditions and two unknowns, we can find a unique solution that tells us what happens in the collision (see Appendix J). The solution shows that the alpha particle will bounce backward only if its mass is less than that of the atomic particle, and will continue its forward motion if its mass is greater than that of the atomic particle. This is because, at the dividing point between the two cases of an alpha particle bouncing back and continuing forward, the alpha particle would have to be just brought to rest, so all its momentum and kinetic energy is given to the struck atomic particle. Momentum and kinetic energy are given by different formulas (one is mass times velocity, the other is one-half the mass times the square of the velocity), so, in order for both the momentum and the kinetic energy of the initial alpha particle and the final atomic particle to be equal in the special case where the alpha particle is just brought to rest in the collision, the masses as well as the initial and final velocities (respectively) of the alpha particle and atomic particle would have to be equal. Geiger and Marsden had observed particles bouncing almost straight back from the gold foil, so Rutherford could conclude that it must be striking some particle at least about as heavy as itself.

Electrons are about 8,000 times lighter than alpha particles, so they are quite ruled out as the atomic particle responsible for the large-angle scattering.

With the benefit of our present knowledge of the nature of the atom, I have "stacked the deck" in describing the problem of the large-angle scattering of alpha particles. As I have explained, these scatterings must be due to encounters with charged atomic particles that are much smaller than an atom and at least as heavy as an alpha particle. We have also seen that there must be some positive electric charges inside the atom to compensate for the negative charges of the electrons, and that there must be something in the atom much heavier than an electron to account for the mass of the atom. Finally, atoms must be mostly empty space, as shown by Lenard's observation that cathode rays can travel great distances in gases. What could be more natural than to suppose that the atom has a small central core or nucleus, which contains most of the mass of the atom, and carries a positive electric charge that attracts the negative electrons and keeps them in orbit around the nucleus?

This totting up of evidence gives an altogether misleading impression of the easiness of Rutherford's task in explaining the large-angle scattering. A great many wrong explanations must have passed through his mind. Perhaps the alpha particle is scattered not by single atoms or subatomic particles, but as a result of interactions with a sizable part of the gold foil. Perhaps the alpha particle is scattered by electrons in the atom, but by electrons that happen to be traveling at enormous speed toward the incoming alpha particle. Perhaps the forces responsible for the scattering have nothing to do with electric attraction or repulsion. Perhaps momentum and energy are not conserved inside atoms. And so on. We have no idea of the sorts of explanation that Rutherford may have briefly considered and then rejected. (Scientists usually try to avoid publishing ideas that turn out not to work.) What we do know is that Rutherford had by 1911 focused on the idea that the atom consists of a small, massive, positively charged nucleus surrounded by orbiting electrons. Geiger recalled that, early in 1911, "one day Rutherford, obviously in the best of spirits, came into my room and told me that he now knew what the atom looked like and how to explain the large deflections of alpha particles."[5] Rutherford had fastened on the idea of the atomic nucleus.

Rutherford's conclusions were announced in a paper read to the Manchester Literary and Philosophical Society on March 7, 1911.[6] By a happy chance, this was the same forum to which Dalton in the early 1800s had communicated his results on atomic weights. Only an abstract survives of Rutherford's talk, but later in 1911 he submitted to the *Philosophical Magazine* a long article, "The Scattering of α and β Particles by Matter and the Structure of the Atom," in which his work is described in detail.[7] The important thing about Rutherford's work is not just that he had gotten the right idea—that an atom consists of a small, heavy, positively charged nucleus surrounded by orbiting electrons—but that he had found a way to test it.

The analysis Rutherford used is one that has been repeated countless times since 1911 in studies of the structure of atoms, nuclei, and elementary particles. Suppose we want to test some hypothesis about the nature of the atom, such as Rutherford's picture of a tiny positive nucleus surrounded by a cloud of electrons. Using this hypothesis together with Newtonian mechanics, we can calculate the hyperbolic orbit of an alpha particle fired at the atom, in much the same way that an astronomer calculates the hyperbolic orbit of a comet passing through the solar system.* Of course, one cannot see into the atom, so the interesting thing is the one thing that can be measured: the scattering angle, the angle between the initial direction of the alpha particle as it comes in from infinity and the direction along which it recedes to infinity again after the encounter. But unfortunately this scattering angle is not fixed; it depends on the line along which the alpha particle approaches the atom. It is convenient to express this dependence in terms of the *impact parameter*, the distance by which the alpha particle would miss the center of the nucleus if it were not deflected. For instance, for an alpha particle with velocity 2.09×10^7 meters per second approaching a nucleus with an electric charge of Z electronic charges, the scattering angle for an impact parameter of 1.5×10^{-16} meters can be calculated to be 90°. (See Appendix for the formula for carrying out such calculations. Incidentally, it is no coincidence that the impact parameter that gives a large scattering angle like 90° is of the same order of magnitude as the distance of closest approach for a head-on encounter that we calculated earlier. In both of these situations the alpha particle gets close enough to the nucleus that its initial kinetic energy is largely used up in doing work against the electrical repulsion of the nucleus, a necessary condition if it is to be strongly deflected.)

How can we possibly use the results of such calculations to analyze experimental data? After all, the alpha particles are not aimed at specific atoms, but are just fired blindly into a foil containing a vast number of invisible atoms. The answer found by Rutherford is that the analysis must be done statistically, not by measuring the scattering angle for a single alpha particle encountering an atom at a known impact parameter, but by measuring the distribution of scattering angles for many alpha particles that happen to pass close to one atom or another at random impact parameters.

---

* But see the footnote on p. 133.

For instance, suppose we measure the fraction of all alpha particles scattered by at least a given angle, say 1° or 90° or 179° or whatever. In order for this to happen, the impact parameter would have to be less than a certain amount; in the example above, it would have to be less than $1.5 \times Z \times 10^{-16}$ meters for the alpha particle to be scattered by at least 90°. For the purposes of calculating the fraction of alpha particles scattered by at least a given angle, each nucleus can be thought of as a little disc facing the incoming alpha particle, the disc radius being the maximum impact parameter for such scattering: Only those alpha particles that happen to hit one of these discs will be scattered by at least the given angle. The fraction of alpha particles scattered by at least the given angle is thus simply equal to the fraction of the area of the foil occupied by these discs—in other words, by the area of each disc times the average number of atoms per unit area in the foil.

By the familiar formula for the area of a circle, the area of each disc is π times the square of the maximum impact parameter for scattering by at least the given angle. This area depends on the scattering angle we are interested in. It is clearly not the actual area of any physical disc, but it is the fundamental quantity that determines the probabilities of scattering by various angles, and is therefore called the *effective cross section* of the atom. A good deal of modern physics consists of the measurement of such cross sections.

For example, we saw that the maximum impact parameter for scattering of an alpha particle by at least 90° in the Geiger-Marsden experiments was calculated to be $1.5 \times Z \times 10^{-16}$ meters. (Recall that Z is the charge of the nucleus in units of the charge of the electron.) The effective cross section was therefore

$$\pi \times (1.5 \times Z \times 10^{-16} \text{ m})^2 = 7 \times Z^2 \times 10^{-32} \text{ m}^2.$$

Also, the number of gold atoms in a square meter of foil is calculated by taking the mass per square meter of foil, which is the density $1.93 \times 10^4 \text{ kg/m}^3$ of gold times the thickness 4×10^{-7} m of the foil, and dividing this by the mass of one gold atom, which is the atomic weight 197 of gold times the mass 1.7×10^{-27} kg of a unit atomic weight. This gives

$$\frac{(1.93 \times 10^4 \text{ kg/m}^3) \times (4 \times 10^{-7} \text{ m})}{197 \times (1.7 \times 10^{-27} \text{ kg})} = 2.3 \times 10^{22} \text{ atoms/m}^2.$$

In one square meter of foil the total area occupied by our fictitious little discs is then the number 2.3×10^{22} of atoms times the area $7 \times Z^2 \times 10^{-32}$ m^2 of each disc, or $1.6 \times 10^{-9} Z^2$ square meters, so the probability that the alpha particle happens to be aimed at one of these fictitious discs and will therefore be scattered by at least $90°$ is $1.6 \times 10^{-9}Z^2$. (The fact that this is much less than 1 shows that we can ignore the possibility of some discs overlapping.) In comparison, Geiger and Marsden had measured this probability to be about one in 20,000, or 5×10^{-5}, so it was possible to conclude that Z, the electric charge of the nucleus, must be approximately

$$Z \approx \sqrt{\frac{5 \times 10^{-5}}{1.6 \times 10^{-9}}} = 180.$$

This is not too good a value; we now know that the nucleus of the gold atom has an electric charge of 79 electronic units. However, Geiger and Marsden had not in 1909 aimed at a precise measurement of scattering probabilities, so the discrepancy is not surprising. Rutherford in his 1911 paper actually used somewhat more precise data of Geiger and Marsden on small-angle scattering of alpha particles, and found values for the nuclear charge Z of gold of 97 in one case and 114 in another. He also used data of J. A. Crowther on the scattering of beta rays to determine Z for various other elements. Table 4.1 shows his results in comparison with the modern values. I do not know why Rutherford's results for Z were systematically too high, but at least they were of the right order of magnitude, and showed that nuclear charge increases with atomic weight, as might have been expected.

Much more important than these crude measurements of nuclear charge was the verification of Rutherford's basic assumption that the scattering is due to a small, heavy, charged nucleus. Rutherford had calculated the impact parameter that would give scattering by a given angle; squaring this and multiplying by π then gave the effective cross section for scattering by that

Table 4.1. Rutherford's calculations of atomic numbers.

| Element | Atomic weight | Nuclear charge Z in units of electron charge | |
		as deduced by Rutherford	as known today
Aluminum	27	22	13
Copper	63	42	29
Silver	108	78	47
Platinum	194	138	78

angle or more.* For instance, according to Rutherford's formula, the effective cross section for scattering by at least 135° is less than that for scattering by at least 90° by a factor 0.00196. As we have seen, the fraction of alpha particles scattered by various angles are just given by the product of these cross sections times the number of atoms per unit area of foil. Starting in 1911, Geiger and Marsden began a program of carefully measuring the fraction of alpha particles scattered by various angles, and in 1913 they reported that their experimental results were in good agreement with Rutherford's theoretical formulas.[8] Thus, the correctness of Rutherford's picture of an atomic nucleus surrounded by electrons was now definitely established.

Atomic Numbers and Radioactive Series

The discovery of the atomic nucleus had one immediate consequence of enormous importance. A few months after Rutherford's paper announcing this discovery appeared in print, on a visit to Cambridge he met Niels Bohr, who then visited Rutherford at Manchester a year later. Bohr seized on the problem of explaining the dynamics of the electrons in their orbits around the nucleus and the emission and absorption of light when electrons make transitions from one orbit to another. His theory was based on the ideas of quantum theory, which lie outside the scope of this book. For our present purposes, just one point is essential: Bohr derived a formula that gave the length of the light waves (usually x rays) emitted when an electron enters one of the innermost orbits of an atom in terms of (among other, known quantities) the electrical charge of the nucleus. Hence, the wavelengths of these x rays could be used to measure the one crucial unknown quantity in Rutherford's picture of the atom, the nuclear charge.

At just this time, a young physicist at Manchester, H. G. J. Moseley (1887–1915), was learning how to measure the wavelengths of x rays with great precision, using crystals in place of diffraction gratings to produce a

---

* Rutherford was lucky to get the right answer for the relation between impact parameter and scattering angle. In general such calculations have to be carried out by the methods of quantum mechanics, and for the energies and masses typical of nuclear physics the results are very different from those that would be obtained by Rutherford's approach, in which one calculates the orbits of the scattered particles by the rules of classical Newtonian mechanics. It happens that there is just one case for which the quantum and the classical approaches give precisely the same answer for scattering problems: the case of forces that decrease as the inverse square of the distance, which of course is just the case that was of interest to Rutherford. If Thomson's "plum pudding" model of the atom had been correct, then classical calculations like those of Rutherford would have given the wrong answer for the cross section, and it would have been impossible to interpret the results of the Geiger-Marsden experiment correctly until the development of quantum mechanics.

wavelength-dependent bending of the rays. After the appearance of Bohr's 1913 papers,[9] Moseley set out to measure the nuclear charges of a series of elements of medium atomic weight that emit x rays in a convenient range of wavelengths. His results, as published in 1913,[10] are shown in Table 4.2.

Several features of the list in Table 4.2 stand out dramatically. First, the nuclear charges are within a small fraction of a percent (which could be attributed to experimental error) of a whole-number multiple, 20, ?, 22, 23, . . ., 30, of the charge of the electron. This in itself was no surprise, since the nuclear charge was supposed to cancel the charge of whatever whole number of electrons the atom contains, so that the atom can be electrically neutral. Nevertheless, the fact that the charge came out as a whole-number multiple of the electron charge did much to reinforce Moseley's faith in his own measurements and in Bohr's theory.

What was not expected was that the nuclear charge simply goes up by one unit as we go from one element to the element next higher in atomic weight. (Cobalt, a minor exception, is understood today in terms of the especially strong binding of the nuclei of the adjacent elements, iron and nickel.) In fact, as Moseley recognized, this pattern extends beyond the elements he studied directly. If all the chemical elements, starting with hydrogen, are listed in a sequence of increasing atomic weight—hydrogen, helium, lithium, and so on, as in the table on pp. 195–196—then calcium is number 20 in the list, titanium number 22, and so on up to zinc, which is number 30, in nearly perfect correspondence with the nuclear charges measured by Moseley. Thus, with a few exceptions, the number that gives the place of an element in the list

Table 4.2. Moseley's measurements of atomic numbers.

Element	Nuclear charge (*in units of electron charge*)	Atomic weight
Calcium	20.00	40.09
Scandium	not measured	44.1
Titanium	21.99	48.1
Vanadium	22.96	51.06
Chromium	23.98	52.0
Manganese	24.99	54.93
Iron	25.99	55.85
Cobalt	27.00	58.97
Nickel	28.04	58.68
Copper	29.01	63.57
Zinc	30.01	65.37

Niels Bohr, around 1922.

H. G. J. Moseley, in the Balliol-Trinity laboratory.

of elements when they are ordered by atomic weight is the same as the electrical charge of the nucleus in units of the charge of the electron, a quantity now called the *atomic number*. Clearly, whatever particles give the nucleus its positive charge, the more of them there are the heavier the atom is.

It was now possible to determine the nuclear charge of any element, and by inference the number of electrons in its atom, by just looking at the list of elements in order of increasing atomic weight. For instance, gold is the element with the 79th lowest atomic weight, so its nucleus must have a positive charge equal to that of 79 electrons; to cancel this charge, the gold atom must contain 79 electrons. Even more important, it was now understood that the particular set of elements found on Earth is not some random sampling of an infinitely diverse menu of elements, but comprises essentially all the elements that could exist (aside from elements like those heavier than uranium, whose half-life is so short that they could not have survived to the present). Elements are like the jokes told at the proverbial comedians' banquet. Just as the comedians only need to give the number of the joke to get a laugh from their colleagues, so chemists only need to give the atomic number—1, or 26, or 79—to evoke all the properties associated with hydrogen, or iron, or gold. And although in Moseley's day there were four gaps in the list of nuclear charges, they have now all been filled in by discovery of the missing elements. As Rutherford's old collaborator Soddy wrote, "Moseley, as it were, called the roll of the elements, so that for the first time we could say definitely the number of possible elements between the beginning and the end, and the number that still remained to be found."

Of all the millions of tragic deaths in the First World War, the one most regretted by the world of physics was that of Moseley. At the outbreak of the war, he hurried back to England from the British Association meetings held that year in Australia, and enlisted in the Royal Engineers, as Signals Officer. He was killed in August 1915 during the Gallipoli campaign, in the landings at Suvla Bay.

The work of Rutherford and Moseley continued to bear fruit. Soddy had pointed out in 1911 that when an atom emitted an alpha particle, it always seemed to turn into an atom of the element two places down in the list of elements in order of atomic weights. Also, Soddy, K. Fajans, and A. S. Russell (all sometime Rutherford collaborators) had noted independently in 1913 that when an atom emitted a beta particle it always seemed to turn into an atom of the element one place up in the list. These "displacement laws" were now neatly explained by Moseley's discovery of the relation between atomic number and nuclear charge. Alpha particles carry a charge of +2 electronic units (note that helium is number 2 in the list of elements), so when an atomic nucleus emits an alpha particle it must lose two units of charge. Also, beta particles are electrons, so they naturally have a negative charge of -1 electronic units, and when a nucleus emits a beta particle its positive charge must increase by one unit. And alpha particles have atomic weight 4, and beta particles have negligible atomic weight, so the isotope resulting from emission of an alpha particle has an atomic weight four units less than that of the original isotope, but the isotope resulting from emission of a beta particle has the same atomic weight as the original isotope. This may all seem pretty obvious, but in 1913 knowledge of the nucleus was only two years old, and the displacement laws could already be cited as evidence that the nucleus was the seat of alpha and beta radioactivity.

The displacement laws also made sense of the complicated sequences of radioactive transformations that had with such difficulty been worked out by Rutherford and Soddy at McGill. Let's see how this works for the thorium series. Natural thorium consists mostly of the long-lived isotope ^{232}Th, and thorium has atomic number 90 (that is, its atom weighs about 232 times as much as the hydrogen atom, and has an electric charge of 90 electronic units). It is observed to emit an alpha particle (half-life 1.41×10^{10} years), so the decay product must have atomic weight $232 - 4 = 228$ and atomic number $90 - 2 = 88$. Now, 88 is the atomic number of radium, so we can conclude that ^{232}Th decays into ^{228}Ra. Next, ^{228}Ra is observed to emit a beta particle (half-life 5.77 years), so it turns into an atom with the same atomic weight and with atomic number $88 + 1 = 89$. This is the atomic number of the element actinium, so we see that ^{228}Ra decays into ^{228}Ac. Next, ^{228}Ac suffers another beta decay (half-life 6.13 hours), so the atomic number goes back to 90 (that of thorium), but we now have the lighter isotope ^{228}Th. Next, ^{228}Th emits an alpha particle (half-life 1.913 years), turning into ^{224}Ra. This is Rutherford's "thorium X," which as we now see is actually the great-granddaughter of natural thorium. Next, ^{224}Ra emits an alpha particle and turns into ^{220}Rn, with atomic number $88 - 2 = 86$. This is Rutherford's "thorium emanation." After four more alpha decays and two more beta decays, the atom finally turns into the commonest isotope of lead, ^{208}Pb, and its radioactivity is at last extinguished. The complete thorium and uranium series are shown on page 138. We will consider in the next section why heavy nuclei undergo these complicated sequences of particle emissions.

Since we have been speaking of atomic numbers, which are always whole numbers, this is a good place to come back to atomic weights and to ask why they are not also whole numbers, equal simply to the number of protons (or protons plus neutrons). The essential ingredient in the answer to this old question was provided in 1905 by Albert Einstein (1879–1955) in a pair of the

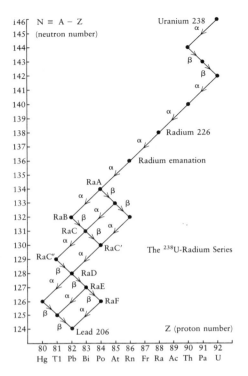

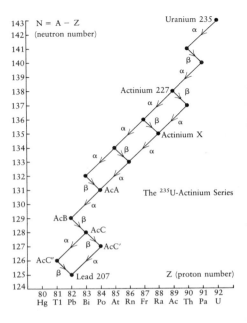

The three principal radioactive series. These figures show the nuclei produced by the three sequences of alpha and beta decays that start with the three very long-lived radioactive isotopes found in the earth: uranium 238, uranium 235, and thorium 232. The horizontal and vertical axes give the atomic number and the difference of atomic weight and atomic number, respectively; the sum of these numbers gives the atomic weight. (Equivalently, the vertical axis gives the number of neutrons; the horizontal axis the number of protons.) Alpha decays, are represented by arrows running from upper right to lower left, beta decays by arrows running from upper left to lower right. Some of the nuclei are labeled with the names given to them in the early history of nuclear physics; for instance, "radium A" is polonium 218, "thorium A" is polonium 216, and "actinium A" is polonium 215. The paths taken by the nuclear sequences shown here mark the general trend of the "stable valley" of nuclei with minimum internal energy.

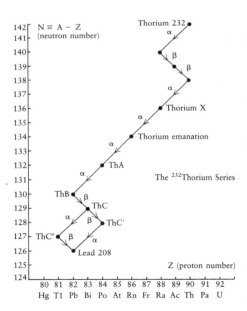

most important papers in the history of physics.[11] The first paper presented Einstein's Special Theory of Relativity, a new understanding of space and time, which lies far outside the scope of this book. The second paper applied Special Relativity to the emission of light by a moving body. Einstein found that the energy released by a moving body is larger than when it is at rest, by an amount proportional to the square of its velocity. His interpretation was that the emission of light not only reduces the energy stored internally in the body, as in the case of a body that emits light when at rest, but also reduces the body's kinetic energy, by decreasing its mass. (Recall that kinetic energy is proportional both to the square of the velocity and to the mass of the moving body.) The general conclusion was that increases or decreases in internal energy are always accompanied by corresponding changes in mass, given by the formula

$$\text{Change in mass} = \frac{\text{Change in internal energy}}{(\text{Speed of light})^2}.$$

This is the original version of the famous formula $E = mc^2$.

Now, the speed of light in ordinary units is a very large number (2.9979×10^8 meters per second), so for most everyday processes these changes in mass are too small to be detected. For instance, it has already been mentioned that the burning of a kilogram of natural gas releases about 5×10^7 joules. After the heat dissipates, the combustion products will be found to weigh less than one kilogram by the amount

$$\frac{5 \times 10^7 \, \text{J}}{(3 \times 10^8 \, \text{m/sec})^2} = 5.5 \times 10^{-10} \, \text{kg},$$

which is less than the mass of a speck of dust. Einstein knew about the much larger energies released in radioactivity, and he speculated that "it is not impossible that with bodies whose energy content is variable to a high degree (e.g. with radium salts), the theory may be substantially put to the test."

Einstein was right, but the test would not come until Thomson and Aston began sorting out the different isotopes and precisely measuring their atomic weights. We now know that internal energy does indeed contribute to mass, just as predicted by Einstein's theory. For instance, in the alpha radioactivity of the most common uranium isotope, ^{238}U, an energy of 6.838×10^{-13} joules per nucleus is released, most going into the kinetic energy of the alpha particle. According to Einstein's formula, the mass of the decay products when brought to rest should be less than that of the ^{238}U nucleus by the amount

$$6.838 \times 10^{-13} \text{ J}/(2.9979 \times 10^8 \text{ m/sec})^2 = 7.608 \times 10^{-30} \text{ kg}.$$

A unit atomic weight corresponds to a mass of 1.66×10^{-27} kg,* so we can restate this by saying that the atomic weight of the decay products should be less than that of the ^{238}U nucleus by the amount

$$\frac{7.608 \times 10^{-30} \text{ kg}}{1.66 \times 10^{-27} \text{ kg/amu}} = 0.0046 \text{ amu}.$$

To check this, note that the atomic weight of ^{238}U is 238.0508, and that it decays into an alpha particle with atomic weight 4.0026 and a nucleus of ^{234}Th with atomic weight 234.0436. The mass loss is then

$$238.0508 - 4.0026 - 234.0436 = 0.0046,$$

in perfect agreement with what is expected from Einstein's formula.

  We now see that the atomic weight of an element is not just the number of nuclear particles contained in its nucleus; it also receives a contribution from the internal energy of the nucleus. Thus, it would not be expected to be precisely a whole number. One other reason that the atomic weight is not a whole number is that nuclei consist of differing proportions of two sorts of particles of different mass, originally thought to be protons and electrons and known since the mid-1930s to be protons and neutrons. This turns out to be numerically less important; most of the departure from whole numbers of the atomic weights of all but the lightest nuclei can be attributed to their internal energy, not to the mass differences of their constituents.

  Our line of reasoning can be inverted: From the atomic masses of various isotopes we can judge how much energy can be released in their radioactive decays or other reactions. Looking over a list of atomic weights like Table 3.4 on page 85, we see that the atomic weight is above the nearest whole number for the lightest elements (1.00793 for hydrogen, 4.0026 for helium), equals 12 by definition for carbon, falls below the nearest whole number for medium-weight nuclei (15.99491 for oxygen, 34.96885 for chlorine, 55.9349

---

* This is called an amu (atomic mass unit).

for iron, etc), and then rises back to whole numbers for heavier elements (226.0254 for radium, 232.0382 for thorium, etc.). We can conclude that the internal energy per nuclear particle is lowest for the nuclei of medium atomic weight and larger for both lighter and heavier nuclei. (Why this is the case will be considered in the next section.) Hence, common isotopes of medium-weight nuclei are not radioactive, because they have no surplus of internal energy to get rid of; and common isotopes of light nuclei are not radioactive, because the lighter nuclei into which they might decay have an even larger surplus of internal energy. On the other hand, nuclei of large atomic weight have more surplus energy than the lighter nuclei into which they might decay, so they can and do release this energy in radioactivity.

The Neutron

For twenty years after the discovery of the atomic nucleus, physicists generally thought that the nuclei of all elements consisted of hydrogen nuclei (later called protons) and electrons. Helium has atomic weight 4 and atomic number 2, so its nucleus (the alpha particle) was supposed to consist of four protons and two electrons, to give it a nuclear charge of $4 - 2 = 2$ electronic units. Similarly, a nucleus such as oxygen, with atomic weight 16 and atomic number 8, would be supposed to contain sixteen protons and eight electrons, although it was widely thought that these might be clustered in the form of four alpha particles. And so on up to the heaviest nuclei like uranium, which, with atomic weight 238 and atomic number 92, would have to consist of 238 protons and $238 - 92 = 146$ electrons.

To find out what the nucleus really consists of, it was necessary to break it up and see what came out. Such nuclear disintegration was first accomplished by Rutherford in 1917, while he was still at Manchester. Rutherford is said to have come in late one day for a War Research Committee meeting, explaining, "I have been engaged in experiments which suggest that the atom can be artificially disintegrated. If it is true, it is of far greater importance than a war!"[12]

Rutherford had noted earlier that a metal source coated with the alpha emitter radium C always gives rise to particles that produce scintillations on a zinc-sulfide screen at a distance beyond the range of alpha particles in air. Studying this phenomenon in a magnetic field, Rutherford concluded that the particles responsible for the scintillations were the nuclei of hydrogen, which we now call *protons*. However, he did not know whether these protons were just recoiling nuclei from hydrogen atoms that happened to be present on the

Rutherford's nuclear-disintegration chamber, in which light nuclei were disintegrated by alpha particles.

metal source and were struck by alpha particles, or whether they were actually knocked out of elements heavier than hydrogen. To study the phenomenon, he put a radium C source in an evacuated metal box with a hole covered by a very thin silver plate. The plate would allow the alpha particles to get out and strike a zinc sulfide screen, and yet would keep air out of the box. Rutherford observed the change in the number of scintillations when various metal foils were placed between the silver plate and the zinc sulfide screen, or when various gases were admitted into the box. For the most part, the rate of scintillations decreased in proportion to the stopping power of the foils or gases. However, when dry air was admitted into the box, the scintillation rate went up! By repeating this experiment with all the constituents of air—oxygen, nitrogen, and so on—Rutherford learned that the effect was due to the collisions of alpha particles from the radium C source with the nuclei of nitrogen in the air.

The process Rutherford discovered was the disintegration of the nitrogen nucleus, in which an alpha particle penetrates into the nucleus and knocks out a proton. The reason that this had not been seen long before is very simple: the electrical repulsion between the positively charged alpha particle and a heavy nucleus like that of gold with a positive charge of 79 electron units was just too strong to allow the alpha particle to get close to the nucleus. (As we saw earlier, even in a head-on collision an alpha particle of typical velocity can only get to within $3 \times 79 \times 10^{-15} = 340 \times 10^{-15}$ meters from the center of a nucleus with atomic number 79, but the gold nucleus is now known to have a radius of only about 8×10^{-15} meters.) Nitrogen, on the other hand, has a nuclear charge of only seven electronic units, so the exceptionally energetic alpha particles emitted by radium C could at least get close to the nucleus, and could occasionally hit an outlying proton. In his report of this result in a 1919 paper, Rutherford concluded as follows:

> *From the results so far obtained it is difficult to avoid the conclusion that the long-range atoms resulting from collisions of α particles with nitrogen atoms are not nitrogen atoms but probably atoms of hydrogen, or atoms of mass 2. If this be the case, we must conclude that the nitrogen atom is disintegrated under the intense forces developed in a close collision with a swift α particle, and that the hydrogen atom which is liberated formed a constituent part of the nitrogen nucleus. . . . The results as a whole suggest that, if α particles—or similar particles—of similar energy were available for experiment, we might expect to break down the nuclear structure of many of the lighter atoms.*[13]

Unfortunately, the discovery of protons knocked out of nitrogen nuclei, together with the long-observed emission of electrons by nuclei as beta rays, only tended to confirm the general view that nuclei consisted of protons and electrons. In a famous talk in 1920, his second Bakerian lecture before the Royal Society, Rutherford speculated prophetically about new kinds of atomic nuclei, but he pictured them all as consisting of protons and electrons.[14] One of the hypothetical nuclei about which Rutherford speculated was a "neutron," with atomic weight 1 and electric charge 0, but this was still pictured as a composite of a proton and an electron. It was entirely unclear to anyone why some of the electrons in an atom should be bound in the nucleus while the others revolved in much larger orbits outside the nucleus, but no one had any idea anyway of what sort of force might be operating at the extremely short distances separating particles within a nucleus.

The discovery of a neutral nuclear particle was made in 1932 at the Cavendish Laboratory by James Chadwick (1891–1974). Chadwick had been a student of Rutherford's at Manchester, and after Rutherford's discovery of the disintegration of nitrogen in 1917–18 he had worked with Rutherford on

James Chadwick.

the disintegration of other light elements, such as aluminum, phosphorus, and fluorine. By 1932 Chadwick was already an established figure in physics, a Fellow of the Royal Society, who served as Rutherford's deputy in running the Cavendish Laboratory, and pursued his own research program.

In 1932 Chadwick's attention was captured by a surprising discovery of Irène and Frédéric Joliot-Curie.[15] It had been found a few years earlier by W. Bothe and H. Becker that beryllium and other light elements, when bombarded with the very fast alpha particles from the radioactive element polonium, emitted highly penetrating radiation, much more penetrating than the protons emitted in nuclear disintegrations like those earlier studied by Ruther-

ford. The rays were at first thought to be electromagnetic radiation, like light
or x rays or gamma rays. Then the Joliot-Curies observed that the rays from
beryllium, when directed into a hydrogen-rich substance such as paraffin wax,
would eject protons from the substance. This in itself might not have been so
surprising, but the protons were found (by attempts to deflect them in a magnetic field) to have a remarkably high speed. The Joliot-Curies calculated that
if the rays emitted from beryllium were really electromagnetic radiation, the
beryllium nucleus must be releasing ten times more energy than was carried by
the alpha particle that produced the rays in the first place. The Joliot-Curies
were even led to question whether the law of conservation of energy was being
violated in these processes.

Chadwick began to study the beryllium rays, directing them into various other materials besides paraffin. He soon found that nuclei other than
hydrogen would also recoil when struck with these rays, but that they moved
with a velocity much less than for hydrogen. The pattern of decreasing recoil
velocities with increasing atomic weight of the recoiling nucleus was just what
would be expected if the beryllium ray was not electromagnetic radiation but a

Chadwick's neutron chamber.

particle with a mass close to that of the proton. Just as in the collisions of alpha particles with nuclei, in a head-on collision with given masses and a given velocity of the particles of the beryllium rays there are two unknowns: the final velocity of the ray particles and the recoil velocity of the nucleus they strike. There are also two conditions constraining them: the conservation of energy and the conservation of momentum. It is therefore possible to solve for both unknown velocities (see Appendix J). In particular, one finds that the recoil velocity of the struck nucleus is given by the following formula:

$$\begin{matrix} \text{Recoil velocity} \\ \text{of struck nucleus} \end{matrix} = 2 \times \begin{matrix} \text{Initial velocity} \\ \text{of ray particle} \end{matrix} \times \frac{\text{Atomic weight of ray particle}}{\begin{matrix} \text{Atomic weight of nucleus} \\ \text{plus atomic weight of ray particle} \end{matrix}}.$$

The initial velocity of the ray particle was not known, but by taking the ratio of the recoil velocities for two different struck nuclei one could eliminate it from the problem and then solve for the atomic weight of the ray particle. For instance, Chadwick (using data of Norman Feather) observed that the same beryllium ray that caused hydrogen nuclei (atomic weight 1) to recoil with a velocity of 3.3×10^7 m/sec would cause nitrogen nuclei (atomic weight 14) to recoil at 4.7×10^6 m/sec. For a fixed initial velocity and atomic weight of the ray particle, the above formula shows that these recoil velocities are just inversely proportional to the sum of the atomic weight of the ray particle and that of the struck nucleus, so

$$\frac{3.3 \times 10^7}{4.7 \times 10^6} = \frac{14 + \text{Atomic weight of ray particle}}{1 + \text{Atomic weight of ray particle}}.$$

The solution is that the atomic weight of the ray particle emitted when beryllium is struck by an energetic alpha particle is 1.16, because then the right-hand side has the value $15.16/2.16 = 7.02$, which is also the value of the left-hand side. Unfortunately, the velocities here were not known with a precision of better than about 10 percent, so Chadwick concluded only that the mass of the ray particle would have to be very nearly equal to the mass of the hydrogen nucleus, the proton.

One other property of the beryllium ray particles was clear from the start: their great penetrating power meant that they must be electrically neutral. (Charged particles are deflected by the electric fields within atoms; this is why the electrically neutral gamma rays are much more highly penetrating than alpha or beta rays.) It seemed, then, from its atomic weight and its neutrality, that the particle produced by alpha rays in beryllium was just the electrically neutral composite of a proton and an electron about which Rutherford had speculated in his Bakerian lecture in 1920. Chadwick reported this result to the Kapitza Club, an informal circle of physicists that had been brought together at the Cavendish by the Russian physicist Peter Leonidovich Kapitza (b. 1894). A few days later Chadwick published the discovery in *Nature* (February 27, 1932), and more fully a little later that year in the *Proceedings of the Royal Society*.[16] In the latter report, Chadwick called this particle by the name by which it has since been known: the neutron.

For Chadwick, as for Rutherford, the neutron was merely a composite of a proton and an electron, not an elementary particle in its own right. This view was reinforced by a more precise measurement of its mass (using neutrons from boron instead of beryllium), which seemed to indicate that the neutron's mass was slightly less than the mass of the proton plus the mass of the electron, as would be expected on the basis of Einstein's relation between energy and mass if the neutron were really such a composite. (The internal energy and hence the mass of a composite system must be less than that of its constituents; otherwise energy could be released by the dissolution of the composite into its constituents, and it would therefore be unstable.)

Chadwick did not speculate in his 1932 papers about the role of the neutron in the structure of the nucleus. This problem was taken up immediately by the German theorist Werner Heisenberg (1901–1976), already famous as one of the pioneers of quantum mechanics in 1925–26.* In a series of 1932 papers in the *Zeitschrift für Physik*,[17] Heisenberg proposed that nuclei consist of protons and neutrons and are held together by the exchange of electrons between them. That is, a neutron gives up its electron and becomes a proton, and the electron is then picked up by another proton which becomes a neutron. Energy and momentum as well as charge are exchanged here, giving rise to what is called an *exchange force*. However, since the neutron was still thought of by Heisenberg (at least in this connection) as a composite of a proton and an electron, the nucleus could still be regarded as built up ultimately from protons and electrons.

---

* Other physicists besides Heisenberg began at this time to consider the idea that the nucleus is made up of protons and neutrons. At a symposium on *Nuclear Physics in Retrospect,* Emilio Segrè mentioned among them the Soviet physicists D. Iwanenko and I. E. Tamm and the Italian Ettore Majorana, who disappeared mysteriously a few years later after a short but brilliant career. Segrè recalls that he was with Majorana when they first heard the news of the discovery by the Joliot-Curies of the fast proton recoils produced by the penetrating radiation from beryllium. Majorana exclaimed, "Oh, look at the idiots; they have discovered the neutral proton, and they don't even recognize it."

This view of the nucleus had already been contradicted, and the contradiction came from a surprising source. In 1929 Walter Heitler (b. 1904) and Gerhard Herzberg (b. 1904) had pointed out that the spectra of diatomic molecules such as oxygen (O_2) or nitrogen (N_2) depended critically on whether their atomic nuclei contained an odd or an even number of elementary particles, then thought to be protons and electrons. Molecules, like atoms, can occupy only certain states of definite energy, and their spectra are produced when light is emitted or absorbed in transitions between these energy levels. In a molecule with two identical nuclei, each containing an even number of elementary particles, half the molecular energy levels that would normally be present in a pair of nonidentical nuclei are absent. If the nuclei are identical but each contains an odd number of particles, then the other half of the energy levels are absent. On this basis, it was found that the oxygen nucleus contains an even number of particles. This was no surprise. With atomic weight 16 and atomic number 8, the oxygen nucleus was considered to consist of 16 protons and $16 - 8 = 8$ electrons, giving $16 + 8 = 24$ particles in all—an even number. The surprise came when Heitler and Herzberg found (on the basis of measurements by F. Rasetti) that the nitrogen nucleus also contains an even number of particles. Nitrogen has atomic weight 14 and atomic number 7, so if its nucleus consists only of protons and electrons there would have to be 14 protons and $14 - 7 = 7$ electrons, making $14 + 7 = 21$ particles in all—an odd number, contradicting the results from the N_2 molecular spectrum.

The solution was to suppose that the neutron is an elementary particle, like the proton and the electron. If one supposed that the nucleus consists of protons and neutrons, then, since neutrons have about the same mass as protons, the atomic weight (rounded off to the nearest whole number) would have to equal the total number of neutrons plus protons, while the atomic number would just equal the number of protons, since they are the only charged particles in the nucleus. That is, the numbers of protons and neutrons are given by the rules

$$\text{Number of protons} = \text{Atomic number}$$

and

$$\text{Number of neutrons} = \text{Atomic weight minus atomic number},$$

so that their sum is the atomic weight. Thus, the ^{16}O nucleus would contain eight protons and eight neutrons, or sixteen particles in all, still an even number. On the other hand, ^{14}N would consist of seven protons and seven neutrons, making $7 + 7 = 14$ particles in all, an even number, in agreement with the evidence from molecular spectra.

Chadwick knew about this line of reasoning, but he does not seem to have taken it very seriously. Near the end of his 1932 paper, he remarked, "It is, of course, possible to suppose that the neutron is an elementary particle. This view has little to recommend it at present, except the possibility of explaining the statistics of such nuclei as N^{14}." (Chadwick used the word *statistics* here because the distinction between nuclei containing odd or even numbers of elementary particles also determines the behavior of large numbers of such nuclei, as described by statistical mechanics.) I do not know why Chadwick and some others paid so little attention to the problem of the molecular spectra, except that there seems to have been a disinclination to introduce new elementary particles—a disinclination so powerful that physicists would rather consider giving up well-established physical principles than contemplate a new particle. We saw one example of this earlier in this section, when the Joliot-Curies were willing to consider giving up the conservation of energy rather than postulate a new massive neutral particle to explain the behavior of the beryllium rays (they were not familiar with Rutherford's 1920 idea of a bound electron-proton pair), and we shall see two other examples when we come to the neutrino and the positron in the next chapter.

It is difficult to pinpoint the moment at which the neutron became accepted as a fully accredited elementary particle. One influence was a more accurate measurement of the neutron mass. By using gamma rays to break up the ^2H nucleus (the *deuteron*) into a proton and a neutron, Chadwick and Maurice Goldhaber (b. 1911) found in 1934 that the neutron mass was a little *larger* than that of a proton plus an electron—not what one would expect if it were a proton-electron composite. (Its mass is now known to be 0.138 percent higher than that of a proton and 0.083 percent higher than that of a proton plus an electron.) Perhaps the most influential experiment was one done in the United States in 1936, by Merle A. Tuve (1901–1982) with N. Heydenberg and L. R. Hafstad, on the scattering of protons by protons.[18] According to the ideas of Heisenberg, protons and neutrons can exert forces on each other by exchanging electrons, but protons do not contain electrons, so there should be no force between them except of course for the much weaker force of electrical repulsion. Tuve, Heydenberg, and Hafstad found instead that protons are scattered strongly in a hydrogen target (by protons), which indicated that the force between two protons is just about as strong as the force between a proton and a neutron. In a companion paper Gregory Breit and Eugene Feenberg proposed that nuclear forces are charge-independent: they behave as if the neutron and the proton are twin brothers.[19] (A similar suggestion was made in the same

The million-volt Van de Graaff accelerator used for the proton-proton scattering experiment. Shown from left to right are O. Dahl, C. F. Brown, L. R. Hafstad, and M. A. Tuve, in 1935.

issue of the *Physical Review* by B. Cassen and E. U. Condon.) It was no longer possible to suppose that the neutron is any less elementary than the proton.

If neutrons are not composed of protons and electrons, and if there are no other electrons in the nucleus, then how are we to understand the fact that electrons are emitted by nuclei in beta radioactivity? The answer was provided in 1933, the year after the neutron was discovered, in a new theory of beta radioactivity developed in Rome by Enrico Fermi (1901–1954).[20] (It is painful to note that Fermi's paper was rejected by the journal *Nature* when it was first submitted there.) In Fermi's theory the emission of an electron in beta radioactivity is just like the emission of light by an excited atom—neither the beta

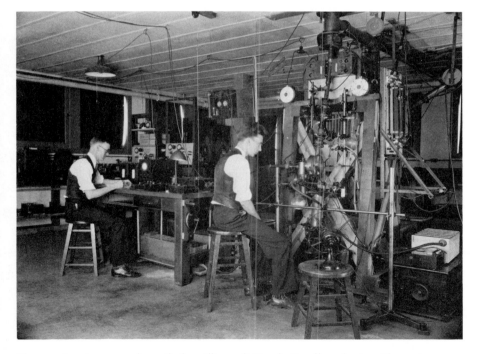

The experiment room underneath the million-volt Van de Graaff accelerator. The proton beam, which was accelerated from an ion source located in the oblong terminal through the long glass accelerator tube (seen in the above figure), entered this room through the ceiling, was deflected by an electromagnet to remove particles not protons, and ended in the small scattering chamber at which Heydenberg (center) is staring.

particle nor the light is "in" the atom until the moment it is emitted—but the emission of beta particles is due not to electromagnetism, but to an entirely new class of force that has come to be known as the *weak interaction*. In a lovely metaphor, George Gamow once compared beta radioactivity to the blowing of soap bubbles: The electron is not in the nucleus before it is emitted any more than the bubble is in the bubble pipe before it is blown.

Chadwick's discovery of the neutron, together with Fermi's theory of beta radioactivity and the accelerators of Cockcroft and Walton and E. O. Lawrence, opened up the modern era of nuclear physics. Most of the subsequent work on the nucleus is outside the scope of this book, but it may be interesting to see how it became possible after 1933 to understand the patterns of radioactive alpha and beta decays that had been worked out experimentally by Rutherford and Soddy much earlier at McGill.

Suppose we were to make a map of all isotopes of all the elements, using the number of protons and the number of neutrons as coordinates instead of longitude and latitude, and suppose we were to mark on this map contours that label values of the nuclear energy per nuclear particle. Starting with the light nuclei, we would see a deep valley running diagonally across the map from lower left to upper right. The nuclei on the valley floor would be those with equal numbers of protons and neutrons: ^{4}He, ^{6}Li, ^{8}Be, ^{10}B, ^{12}C, ^{14}N, ^{16}O, and so on. For complicated reasons, nuclear forces give these nuclei an especially strong binding and hence an especially low energy. Nuclei on the neutron-rich side of this valley have higher energy than the nuclei with equal numbers of neutrons and protons, so energy will be released if a neutron turns into a proton. If enough energy is available to make an electron to balance the charge, then the transitions will occur and the nucleus will exhibit beta radioactivity. For instance, the well-known isotope ^{14}C (eight neutrons, six protons), which is made in our atmosphere by cosmic rays, is neutron-rich, and therefore emits an electron and turns back into the commonest nitrogen isotope, ^{14}N (seven neutrons, seven protons). The neutron itself also suffers beta decay into a proton with a half-life of about 15 minutes, but this was not observed until 1948. Nuclei that are far over on the proton-rich side of the stable valley also exhibit a kind of beta decay; we will come back to this in Chapter 5.

Following the stable valley up toward heavier elements, we would find that it grows steadily deeper, because the attraction caused by the nuclear forces increases with the increasing numbers of protons and neutrons. For this reason, energy will be released if the nuclei of light elements fuse to form heavier nuclei; this is the source of the energy of stars like our sun. Then, for nuclei with more than about twenty protons, a new factor enters. For light

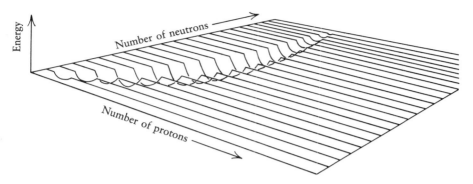

Schematic view of nuclear energies plotted against the number of protons and neutrons in the nucleus, showing the "stable valley" of the more deeply bound nuclei.

nuclei the repulsive electrical forces among the protons are much weaker than the very strong nuclear forces among the neutrons and protons, but the electrical forces build up for larger nuclei much more rapidly than the nuclear forces do, and they become important when there are more than about twenty protons. This fact has two effects: The floor of the stable valley starts to rise again; and the valley veers toward the neutron-rich side of the map, since it is the charge of the protons that is responsible for the rise in the nuclear energy. Although the valley floor is rising, the rise is gentle, and for elements of medium atomic weight there is not enough energy increase as we go from one atomic weight to another four units higher to allow the heavier nucleus to decay into the lighter one by emitting an alpha particle. For elements heavier than lead, the valley floor rises more steeply; and here enough energy is available that the heavier nuclei can get rid of their extra charge by emitting alpha particles. However, emission of alpha particles does not change the excess of neutrons over protons (alpha particles contain two of each), so when a heavy nucleus turns into a lighter one by emitting an alpha particle the new nucleus will have a neutron excess that would be appropriate for a heavier nucleus. The valley of nuclear energies turns steadily toward larger neutron excesses as we go up in atomic weight, so the nuclei produced in alpha decay will typically be on the neutron-rich side of the valley. After another alpha decay, the remaining nucleus will be even more neutron-rich. Eventually, after enough alpha decays (sometimes only one is needed), the remaining nucleus will be so far from the valley that enough energy will be available to create an electron, and a beta decay will occur, moving the nucleus back toward the valley floor. The pattern, then, is of a sequence of alpha decays interspersed with beta decays, the alpha decays taking the nucleus down the valley toward lead but also somewhat out of the valley toward the neutron-rich side and the beta decays returning the nucleus to the neighborhood of the valley floor. This pattern is clearly visible in the radioactive series pictured on p. 136.

We can now come back to the question of how the energy released in radioactivity got into the nuclei in the first place. The universe is believed to have started in a "big bang," after which the primordial hot gas of free protons and neutrons cooled rapidly and at the end of the first three minutes combined into hydrogen and helium. Hydrogen nuclei have a much higher energy per nuclear particle than helium nuclei, and helium nuclei have a higher energy per nuclear particle than those of medium atomic weight; thus, when stars form, the hydrogen nuclei fuse into helium nuclei, and helium nuclei fuse into medium-weight nuclei, releasing enough energy to keep stars shining for billions of years. Eventually the material of a star evolves into those elements around iron whose nuclei have the lowest energy per nuclear particle. There is no more energy to release, and the star begins to cool. Often the star winds up as a cinder, a black dwarf. Sometimes, however, it becomes unstable, begins to implode under the influence of gravity, then may explode as what astronomers call a *supernova*. During such an explosion an intense flux of neutrons is released from the inner part of the star. The neutrons that strike nuclei of medium atomic weight in the outer layers of the star rapidly build them up into heavier elements, all the way up to uranium. The exploding star sheds its outer layers, which then go out to form part of the interstellar medium, out of which a later generation of stars like the sun eventually form. According to this picture, the energy in naturally radioactive elements such as thorium and uranium was put into them by the neutrons released in the explosions of stars, and can ultimately be traced to the force of gravitational attraction, which provided the energy for the stellar explosions.

In recent years the neutron has acquired an ominous practical significance. Neutrons carry no electrical charge, so they are unaffected by the powerful electrical fields near a nucleus that repel alpha particles and other nuclei. Therefore, as Rutherford pointed out in his 1920 Bakerian lecture, they should be able easily to penetrate into even heavy nuclei and cause nuclear disintegrations. It was discovered in 1938 by Otto Hahn (1879–1968) and Fritz Strassmann (b. 1902) that neutrons can cause heavy nuclei to undergo fission.[21] Each fission can produce more than one neutron, so a nuclear chain reaction becomes possible. It is not yet clear whether we will learn how to live with this discovery.

Notes

1. E. Rutherford and F. Soddy, "The Cause and Nature of Radioactivity," *Philosophical Magazine* Series 6, 4 (1903), 561, 576.

2. E. Rutherford and F. Soddy, "Radioactive Change", *Philosophical Magazine* Series 6, 5 (1904), 576.

3. H. Geiger, "On a Diffuse Reflection of the α-Particles," *Proceedings of the Royal Society* **A82** (1909), 445.

4. Quoted by E. N. da Costa Andrade, *Rutherford and the Nature of the Atom* (Doubleday, Garden City, N.Y., 1964).

5. *Ibid.*

6. E. Rutherford, "The Scattering of the α and β Rays and the Structure of the Atom," *Proceedings of the Manchester Literary and Philosophical Society* IV, 55 (1911), 18.

7. E. Rutherford, "The Scattering of α and β Particles by Matter and the Structure of the Atom," *Philosophical Magazine* Series 6, 21, (1911), 669.

8. H. Geiger and E. Marsden, "The Laws of Deflection of α Particles through Large Angles," *Philosophical Magazine* Series 6, 25 (1913), 604.

9. N. Bohr, "On the Constitution of Atoms and Molecules,", *Philosophical Magazine* Series 6, 26 (1913), 1, 476, 857.

10. H. G. J. Moseley, 'The High-Frequency Spectrum of the Elements," *Philosophical Magazine* Series 6, 26 (1913), 257.

11. A. Einstein, "Zur Electrodynamik bewegter Körper," *Annalen der Physik* 17 (1905), 891; "Ist die Trägheit eines Körpers von seinem Energieinhalt abhängig?" *ibid.* 18 (1905), 639.

12. Quoted by N. Feather, *Lord Rutherford* (Priory Press, 1973).

13. E. Rutherford, "Collision of α Particles with Light Atoms IV. An Anomalous Effect in Nitrogen," *Philosophical Magazine* Series 6, 37 (1919), 581.

14. E. Rutherford, "Nuclear Constitution of Atoms," *Proceedings of the Royal Society* A 97 (1920), 374.

15. I. Curie and F. Joliot, *Comptes Rendus Acad. Sci. Paris* 194 (1932), 273.

16. J. Chadwick, "The Existence of a Neutron," *Proceedings of the Royal Society* A 136 (1932), 692.

17. W. Heisenberg, "Structure of Atomic Nuclei," *Zeitschrift für Physik* 77 (1932), 1; 78 (1932), 156; 80 (1932), 587.

18. M. A. Tuve, N. Heydenberg, and L. R. Hafstad, "The Scattering of Protons by Protons," *Physical Review* 50 (1936), 806. Also see G. Breit, E. U. Condon, and R. D. Present, "Theory of Scattering of Protons by Protons," *ibid.* 50 (1936), 825.

19. G. Breit and E. Feenberg, "The Possibility of the Same Form of Specific Interactions for all Nuclear Particles," *Physical Review* 50 (1936), 850.

20. E. Fermi, "Versuch einer Theorie der β-Strahlen," *Zeitschrift für Physik* 88 (1934), 161.

21. O. Hahn and F. Strassmann, "Über den Nachweis und das Verhalten der bei der Bestrahlung des Urans mittels Neutronen entstehenden Erdalkalimetalle," *Die Naturwissenschaften* 27 (1939), 11.

5

More Particles

5

More Particles

The roll of elementary particles is by no means limited to those that make up ordinary atoms: the electron, the proton, and the neutron. As an envoi, I will rapidly review the other particles that have been discovered from the beginning of the twentieth century until the present day.

Photons

In 1905, the same *annus mirabilis* which he developed Special Relativity, Albert Einstein proposed that for some purposes light can be regarded as made up of particles, later called *photons*. The existence of photons was verified experimentally by the results of Millikan's studies of the photoelectric effect in 1914–16, by studies of the scattering of x rays by electrons by Arthur Holly Compton (1892–1962) in 1922–23, and since then in a great variety of other phenomena. Photons have zero mass and zero electric charge, and always travel at the speed of light, so they cannot be contained within atoms.

Neutrinos

Chadwick observed in 1914 that the electron emitted in the beta decay of a radioactive nucleus does not (like an alpha particle or gamma ray) emerge with one definite kinetic energy, but rather with a continuous spectrum of energies, ranging from zero up to a maximum value characteristic of the nucleus that emits it. This was very surprising, because one would have expected the energy of the electron to equal the difference in energies between the initial and final nuclei, and to be a fixed quantity for each specific radioactive element. It was possible that the energy was being shared between the electron and an undetected gamma ray. If so, the total energy released should equal the maximum energy of the beta electrons, the energy they have in those beta decays in which the gamma ray happens to carry off a negligible energy. However, in 1927 C. D. Ellis and W. A. Wooster measured the total heat energy produced by a sample of the beta-radioactive nucleus radium E (^{210}Bi), and found that the energy emitted per nucleus was not equal to the observed maximum energy of the beta electrons but instead to their average energy. After this result was confirmed by L. Meitner and W. Orthmann in 1930, it was clear that a crisis was at hand. No less a figure than Niels Bohr was led to doubt whether the process of beta radioactivity respected the conservation of energy. The correct solution was far less radical. In letters* to friends in 1930, Wolfgang Pauli (1900–1958) proposed that another particle besides the electron is emitted in beta decay and shares the available energy, and that this particle (though electrically neutral) is not a gamma ray but is so highly penetrating that its energy is not converted to heat in experiments like that of Ellis and Wooster. After the discovery of the neutron in 1932, Pauli's hypothetical particle became known as the *neutrino,* or "little neutral one."

The neutrino was incorporated into Fermi's 1933 theory of beta radioactivity; the fundamental process was one in which neutron inside (or outside) a nucleus spontaneously turns into a proton, an electron, and a neutrino. (Strictly speaking, the extra particle emitted in such a beta decay is what later came to be called an antineutrino. Antiparticles are discussed below.) Comparing the distribution of electron energies predicted by Fermi's theory with what was experimentally observed made it possible to conclude that the mass of the neutrino must be very small, much less than the mass of the electron. It is known today that the neutrino mass is no greater than about 10^{-4} electron masses. (There are indications that it may have a mass about half this limit. Neutrinos are now also believed to come in at least three different species, some of which may be heavier than this.)

The Fermi theory also made it possible to calculate the cross section for the absorption of neutrinos in matter. Because the basic interaction is so weak, this cross section comes out to be so small that a neutrino with energy typical of those produced in beta radioactivity would be able to penetrate light years of lead before being absorbed. No wonder they had not contributed to the heat energy measured in the Ellis-Wooster experiment. Neutrinos are enormously difficult to detect, but they are also emitted in enormous numbers in nuclear reactors (through the beta decay of the neutron-rich products of nuclear fission), and in 1955 they were at last observed by Clyde L. Cowan, Jr., and Frederick Reines at the Savannah River reactor. Nowadays, neutrinos are pro-

---

* One of these letters, to participants at an international conference on radioactivity, begins "Dear Radioactive Ladies and Gentlemen."

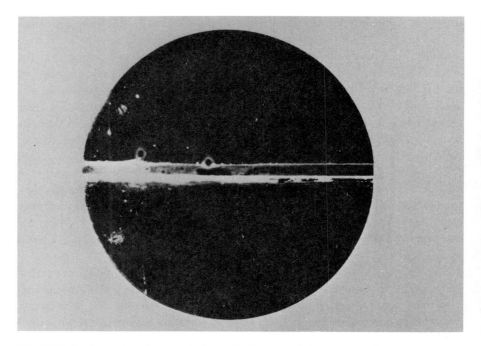

This 1931 cloud-chamber photograph shows the first recorded positron track.

duced in huge numbers from the decay of particles produced in large accelerators, and their interactions have been studied extensively in experiment and in theory. Neutrinos are very common inhabitants of our universe; though cosmic neutrinos have never been observed, there are believed to be about as many of them left over from the big bang as there are photons, and 10^9 to 10^{10} times more of them than protons or neutrons. However, neutrinos interact far too weakly for them to be trapped in the atoms of ordinary matter.

Positrons

The Cambridge theoretical physicist Paul Adrien Maurice Dirac (b. 1902) was engaged in the late 1920s in developing a version of quantum mechanics that would be consistent with special relativity. In the course of this work, he encountered the surprising result that the equation he had worked out to describe a single electron had solutions of negative energy. To explain why there is not a catastrophic collapse of all electrons into these negative energy states, he proposed in 1930 that the states of negative energy are normally already filled, and are thereby unable to accept additional electrons by the same rule (known

Electron-positron pair production. A high-energy gamma ray coming in from above scatters off an atomic electron, losing some of its energy and producing an energetic recoil electron and an electron-positron pair. The electron and positron paths curve because the chamber is placed in a strong magnetic field. The direction of the curves reveals the signs of the particles' charges.

as the Pauli exclusion principle) that keeps the outer electrons in an atom from falling into the inner orbits of lower energy. A few negative energy states might be unoccupied, and these holes in the sea of negatively charged particles with negative energy would appear as particles of positive energy and positive charge. Under the influence of what seems to have been a general code of scientific behavior—that proposing new particles was somehow not respectable—Dirac at first thought that these holes were to be identified with protons. However, Hermann Weyl pointed out that there was a symmetry between holes and electrons, and Dirac was forced to conclude that the holes would have to carry precisely the same mass as electrons. This prediction was unexpectedly verified in 1932 when the American experimentalist Carl Anderson (b. 1905) observed tracks of cosmic-ray particles that curved in a magnetic field about as much as electron tracks, but in the opposite direction. These particles, called *positrons,* are now known with great precision to have the same mass as electrons and the opposite electric charge. They are very rare in the present universe, being produced only in violent astrophysical phenomena like cosmic rays and supernovas and in the rare form of beta radioactivity in which a proton in a proton-rich nucleus turns into a neutron. A positron and an electron that collide are likely to annihilate each other, giving rise to a burst of radiation that carries off the energy in their masses, so positrons are never found in ordinary matter.

Other Antiparticles

After the discovery of the positron, it eventually became clear that to each kind of particle there corresponds an antiparticle, with the same mass as the particle but with opposite values for the electric charge and similar conserved quantities. One essential element in this understanding was a demonstration that antiparticles cannot in general be regarded as holes in a sea of negative-energy particles. In 1934 Pauli and Victor F. Weisskopf showed that even particles that cannot form a stable negative-energy sea have corresponding antiparticles. (Nevertheless, hole theory lingers on in many physics textbooks.) The positron is the antiparticle of the electron, the antineutrino emitted along with electrons in the beta decay of neutron-rich nuclei is the antiparticle of the neutrino that is emitted along with positrons in the beta decay of proton-rich nuclei, and the electrically neutral photon is its own antiparticle. In 1955, Owen Chamberlain (b. 1920) and Emilio Gino Segrè (b. 1905), together with Clyde Wiegand and Tom Ypsilantis, succeeded in producing antiprotons at the Bevatron accelerator in Berkeley. *Antimatter* would consist of antinuclei made up of antiprotons

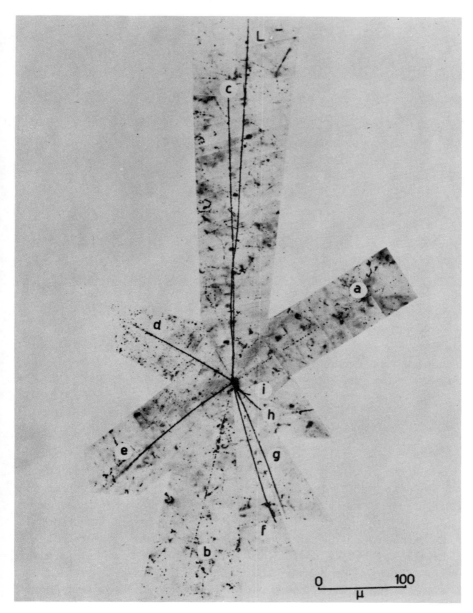

The first antiproton annihilation "star," 1955.

The Bevatron, in the Lawrence Berkeley Laboratory, in 1955.

and antineutrons surrounded by a cloud of positrons. There does not seem to be any appreciable amount of antimatter anywhere within the observable portion of the universe.

Muons and Pions

After the collapse of the idea that nuclear forces are due to the exchange of electrons, the problem remained: What does cause the nuclear force? Could the momentum and energy exchanged in a collision of nuclear particles be carried by some other particle? In 1935 the Japanese theorist Hidekei Yukawa (1907–1981) realized that there is a simple relation between the range of any force and the mass of the particle whose exchange produces the forces: There is a characteristic distance beyond which the force drops rapidly to zero, and this distance is inversely proportional to the mass of the particle whose exchange produces the force. In electromagnetism the exchanged particle is the photon, which has zero mass, so the range of the force is infinite. That is, the force decreases only as the inverse square of the distance, no matter how far. For the exchange of an electron the range of the force comes out to about 10^{-13} meters, and if electron exchange were really the mechanism for nuclear force this is about how large atomic nuclei would be. Actually, atomic nuclei are hundreds of times smaller than this (as was shown by experiments like that of Geiger and Marsden), so this is another mark against the electron-exchange theory of nuclear forces. Yukawa had the courage to propose a new kind of particle, with a mass hundreds of times larger than that of the electron, whose exchange produces a nuclear force with a range of the order of the observed nuclear size, about 10^{-15} meters. Since this particle would be intermediate in mass between electrons and protons, it was called a *meson* (from the Greek *meso*, "middle").

Just two years later, in 1937, a particle with a mass of about 200 electron masses was found in cosmic rays, in cloud-chamber experiments by S. H. Neddermeyer and C. D. Anderson and by C. E. Stevenson and J. C. Street. It was widely assumed at the time that this was the meson predicted by Yukawa. However, an experiment by M. Conversi, E. Pancini, and O. Piccioni in 1945 (while Italy was still under German domination) showed that the mesons that predominate in cosmic rays interact weakly with neutrons and protons—much too weakly for them to provide the mechanism for nuclear force. This puzzle was cleared up by a suggestion (proposed independently by R. E. Marshak and H. A. Bethe in the United States and by S. Sakata and T. Inoue in Japan, and subsequently verified experimentally by C. M. G. Lattes, C. P. S. Occhialini, and C. F. Powell in England) that there are actually two different kinds of mesons, with slightly different masses. The heavier one, now called the pi meson or *pion*, does have strong interactions with protons and neutrons, and contributes to the nuclear force in the manner anticipated by Yukawa; the lighter meson, now called the *muon*, has only weak and electromagnetic interactions, and has nothing to do with Yukawa's theory.

Pi mesons come in three varieties: one that is negatively charged and is 273.1232 times heavier than the electron, its positively charged antiparticle (with precisely the same mass), and a neutral pion that is its own antiparticle and has a mass 264.1129 times that of the electron. This triplet forms a family in the same sense as the proton-neutron doublet; the family relation is required by the symmetry of the nuclear force between protons and neutrons. The muons come in two varieties: a negatively charged particle 206.7686 times as heavy as the electron, and its positively charged antiparticle with the same mass. The muon and the antimuon appear like overweight siblings of the electron and the positron; the difference between them is apparently solely one of mass.

Pions and muons are all unstable: The charged pion and antipion decay respectively into a muon plus antineutrino and an antimuon plus neutrino with an average lifetime of 2.603×10^{-8} seconds, the neutral pion decays into two photons with an average lifetime of about 0.8×10^{-16} seconds, and the muon and antimuon decay respectively into an electron or positron and a neutrino-

One of the first pictures of a pion, 1947. At lower left the pion stops and decays, emitting a muon that travels toward the right.

antineutrino pair with an average lifetime of 2.19712×10^{-6} seconds. The muons that predominate in cosmic rays at sea level are mostly produced by the decay of pions emitted in cosmic-ray collisions with the nuclei of air molecules at high altitudes.

Pions, protons, and neutrons belong to the class of particles known as the *hadrons,* which are distinguished by their participation in the strong nuclear interactions. Muons, electrons and neutrinos belong to the class called the *leptons,* which have no strong interactions but have rather uniform weak (e.g., like beta decay) and electromagnetic interactions. (Another lepton, the *tau,* has recently been discovered at SLAC, the Stanford Linear Accelerator Center.) As far as is known, the similarity in mass between the pions and the muons, which produced so much confusion in the 1930s and 1940s, is essentially a coincidence.

Strange Particles

Physicists might have hoped for a rest after sorting out the pions and the muons, but in the same year, 1947, yet more kinds of particles were discovered in cosmic rays by G. D. Rochester and C. C. Butler. These particles were soon found to be hadrons, in the sense that they participate in the strong interactions, but they appeared strange, because unlike pions they were always produced in pairs. To go into all the properties of all the different kinds of strange particles could take about a hundred times as much space as I have given here to pions, so I will pass over them here.

More Hadrons

The particles I have listed so far are all common in our universe, or at least are abundantly produced by cosmic rays. This aspect of the particle menu changed drastically in the 1950s when large accelerators like the Bevatron at Berkeley and new devices for detecting particles like the bubble chamber began to become available. In the debris of collisions of the high-energy protons from these accelerators were found a great variety of new hadrons, labeled ρ, ω, η, ϕ, Δ, Ξ, Ω, etc.—so many that the Greek alphabet was in danger of exhaustion. They were all unstable, with extremely short lifetimes, which is why they are absent in ordinary matter and had to be produced artificially. However, as far as anyone could tell they were all just as elementary as pions or protons or neutrons. Physicists began to consider a principle, called "nuclear democracy" by Geoffrey Chew at Berkeley, that all the hadrons can be regarded as composites of almost any subset of them, which we can if we like call elementary.

Quarks

Soon an attempt was made to restore some economy to the multitude of hadrons. In the early 1960s Murray Gell-Mann and George Zweig of the California Institute of Technology, building on earlier work of Gell-Mann and Yuval Ne'eman of Tel Aviv, proposed independently that the hadrons are all composites of a few types of really elementary building blocks, called *quarks* by Gell-Mann. Originally there were expected to be just three types of quark, with electric charges $\frac{2}{3}$, $-\frac{1}{3}$, and $-\frac{1}{3}$ in units of the electronic charge, but experiments at SLAC and at Brookhaven have suggested that there are at least two more types, with charges $\frac{2}{3}$ and $-\frac{1}{3}$. Also, each type of quark is believed actually to be a triplet. That makes $5 \times 3 = 15$ quark varieties, and at least one more triplet is expected soon.

It is somehow satisfying that the experimental search for quarks has revived some of the types of experiments that we have been looking at in this book. A 1968 collaboration between SLAC and the Massachusetts Institute of Technology found that high-energy electrons, when fired at protons or neutrons, are sometimes scattered at large angles. This suggested the existence of compact particles within the proton and the neutron, just as the large-angle scattering of alpha particles by gold atoms suggested to Rutherford the existence of a compact nucleus within the gold atom. Also, in the effort to find the quarks, several groups are repeating the experiments of H. A. Wilson and Millikan, hoping to find electric charges that are $\frac{1}{3}$ or $\frac{2}{3}$ the charge of an electron or a positron. Many theorists think that quarks are trapped in the particular combinations we observe as hadrons and can in principle never be observed in isolation, but as always the experimentalists will have the last word.

•   •   •

So it goes. The list of fundamental particles will certainly continue to expand. Physicists are now awaiting the discovery not only of another quark triplet but also of particles called *intermediate vector bosons,* extremely heavy siblings of the photon, and of other particles called *Higgs bosons,* whose properties are less clear.

I returned to the Cavendish Laboratory to talk about these expected new particles and other things in a set of Scott Lectures in 1975. Much had changed since the 1930s. Rutherford, of course, was no longer there to growl at a visiting theorist, as he had when Niels Bohr gave the Scott Lectures almost 50 years before. The present Cavendish Professor, Sir Brian Pippard, is quite kind to theorists. The Cavendish had moved from its old buildings in Free School Lane to a modern complex out of town on the Madingley Road, and the focus of its activity had also shifted, away from nuclear physics and toward radio astronomy, molecular biology, and solid-state physics. But I was very glad to be there. We physicists are always trying to do something new, but we work in an ancient tradition, and we have our own shrines and heroes. The tradition represented by the Cavendish Laboratory is for us as moving as that embodied for followers of other disciplines in the lovely old college buildings along the Cam.

I hope that the reader will not conclude from the account of particle physics I have given here that this branch of physics has degenerated into a kind of butterfly collecting, with the peculiarity that the butterflies we collect do not live long enough to be found in nature and have to be created in the laboratory of the collector. I think that this view is quite wrong. Once the age-old question of the nature of ordinary matter was settled by the discovery of the electron, proton, and neutron, the question shifted. The real task we address in our experimental and theoretical studies of elementary particles is not to develop a list of particles and their properties. It is to understand the underlying principles that dictate why nature—particles, nuclei, atoms, rocks, and stars—is the way it is. All our experience shows that the study of elementary particles is at present the best and perhaps the only way of getting at the fundamental laws of nature.

I also hope that the story told in this book will not create an impression that the history of physics consists in the discovery and study of particles, forces, or any other specific phenomena. Running along with the wonderful discoveries and measurements of Thomson, Becquerel, Rutherford, Millikan, and Chadwick has been an evolution of ideas, a widening of our understanding of physical principles. They have gone together: although I could not go into it here, the discovery of the electron did much to stimulate the development of relativity and quantum mechanics, and in recent years the study of the strong and weak nuclear forces has deepened our understanding of the role of symmetry in nature. But although the discovery of the subatomic particles is not the whole of twentieth-century physics, it is an indispensable part of the story.

When the poet William Blake needed to summarize all of science in one line, he spoke of "the atoms of Democritus, and Newton's particles of light." From the Greece of Democritus and Leucippus to Blake's time and our own, the idea of the fundamental particle has always been emblematic of the deepest aim of science: to understand the complexity of nature in simple terms.

Appendix

A   Newton's Second Law of Motion

In a completely general system of units, Newton's Second Law would state that force is proportional to mass times acceleration; that is,

$$F = kma, \tag{A.1}$$

where F is the force acting on a particle, a is the acceleration given to the particle by this force, m is the particle's mass, and k is a constant whose value depends on the system of units chosen for F, m, and a. It is nearly universal to choose the units of force so that a mass $m = 1$ will be given an acceleration $a = 1$ by a force $F = 1$; for instance, one newton is defined as the force that will give one kilogram an acceleration of one meter/ sec^2. In such a system of units, the constant k must have the value $k = 1$, because otherwise Equation (A.1) would not be satisfied in the special case $m = 1, a = 1, F = 1$. Hence for such units, Newton's Second Law takes the form in which it is most familiar

$$F = ma. \tag{A.2}$$

For instance, on the basis of what we now know about the electron, we can estimate that in Thomson's experiments with cathode rays the force acting on the electron was typically of the order

$$F = 10^{-16} \text{ newton},$$

and the mass of the electron is about

$$m = 9 \times 10^{-31} \text{ kilogram};$$

so that the acceleration was about

$$a = F/m = 1.1 \times 10^{14} \text{ meters/sec}^2.$$

With such an acceleration, after only 10^{-6} seconds the electron would be traveling at 1.1×10^8 meters/sec, a fair fraction of the speed of light (3×10^8 meters/sec). However, the electrons in Thomson's experiment were exposed to the force for only about 10^{-9} seconds, and never came close to light speed.

In the foregoing example, Newton's Second Law was used to calculate the acceleration that a given force would produce in a given mass, but of course it can also be used to calculate the force required to produce a given acceleration of a given mass. For instance, it is a matter of common observation that bodies near the surface of the earth fall with a constant acceleration, 9.8 meters/sec^2, to which is commonly given the symbol g. It follows that the force of gravity on a body of mass m is

$$F_{\text{grav}} = mg, \tag{A.3}$$

whether or not the body is actually free to fall. The force of gravity on an electron is then 9×10^{-31} times 9.8, or 9×10^{-30} newton. This is negligible compared with the electric and magnetic forces on the electron in a cathode-ray tube; so gravity can safely be ignored in analyzing the behavior of electrons in Thomson's experiment.

B Electric and Magnetic Deflection of Cathode Rays

Here we will show how Newton's Second Law is used to calculate the deflection of the cathode ray in Thomson's experiment, and how measurements of this deflection can be used to calculate the mass-to-charge ratio of the ray particles.

Suppose that a force F is exerted on the cathode-ray particles, acting in a direction transverse to the ray's motion. The particles will be given an acceleration in this direction of magnitude $a = F/m$ (where m is the particle mass); so if they are exposed to the force for a time t, they will acquire a component of velocity perpendicular to their original motion, with magnitude

$$v_{\text{perp}} = ta = tF/m. \tag{B.1}$$

Suppose the particles have a component of velocity v in the original direction of the ray, and travel with this velocity through a "deflection region" of length ℓ, where they are exposed to the force F. Since velocity is distance per time, $v = \ell/t$, and therefore the time during which the particles are accelerated is

$$t = \ell/v. \tag{B.2}$$

Substituting this for t in Equation (B.1) gives

$$v_{\text{perp}} = F\ell/mv. \tag{B.3}$$

After leaving the deflection region, the ray particles travel through a "drift region" of length L, in a direction pretty close to the original ray direction, and with a component of velocity in this original direction still equal to v. By the same reasoning that led to (B.2), the time spent in the drift region is

$$T = L/v. \tag{B.4}$$

During this time the ray particles are also moving in a direction perpendicular to their original direction with a velocity v_{perp}; so their displacement from the original path of the ray when they reach the end of the drift region is

$$d = Tv_{\text{perp}}. \tag{B.5}$$

Inserting (B.4) and (B.3) in (B.5) gives

$$d = \left(\frac{L}{v}\right) \times \left(\frac{F\ell}{mv}\right)$$

or, more succinctly,

$$d = \frac{F\ell L}{mv^2}. \tag{B.6}$$

This is the formula quoted on p. 31.

Now let's consider specific kinds of force. If the cathode-ray particles have electric charge e, then the electric force exerted on them by an electric field E is

$$F_{\text{elec}} = eE, \tag{B.7}$$

and according to (B.6), this will produce a displacement of the ray at the end of the tube equal to

$$d_{\text{elec}} = \frac{eE\ell L}{mv^2}. \tag{B.8}$$

The magnetic force exerted by a magnetic field B on a particle of charge e and velocity v (perpendicular to the field) is given by the product of e, v, and B. In Thomson's experiment, v_{perp} was much less than v; so in this case

$$F_{\text{mag}} = evB, \tag{B.9}$$

and the force acts essentially at right angles to the ray's original direction. According to Equation (B.6), this force produces a displacement of the ray at the end of the tube equal to

$$d_{\text{mag}} = \frac{eB\ell L}{mv}. \tag{B.10}$$

Note that the factor v in (B.9) has canceled one of the two factors of v in the denominator of (B.6).

Now, suppose that d_{elec} and d_{mag} are measured for given values of E, B, ℓ, and L. How do we solve for the ratio of the electron's mass and charge? Note that the ratio of (B.10) and (B.8) gives

$$d_{\text{mag}}/d_{\text{elec}} = \frac{eB\ell L/mv}{eE\ell L/mv^2} = \frac{Bv}{E},$$

or, in other words,

$$v = \left(\frac{E}{B}\right)\left(\frac{d_{mag}}{d_{elec}}\right). \qquad \textbf{(B.11)}$$

Inserting this into (B.10) then gives

$$d_{mag} = \frac{eB\ell L}{mEd_{mag}/Bd_{elec}} = \frac{eB^2\ell Ld_{elec}}{mEd_{mag}}$$

Solving for m/e, we now have

$$\frac{m}{e} = \frac{B^2\ell Ld_{elec}}{E(d_{mag})^2}. \qquad \textbf{(B.12)}$$

This is the formula used to deduce the mass-charge ratio of the cathode-ray particles from measurements of their deflection.

For an example, look at the last line of Table 2.1 (p. 54), which gives some of Thomson's 1897 data. The electric and magnetic fields here had the values

$$E = 1.0 \times 10^4 \text{ newtons/coulomb,}$$
$$B = 3.6 \times 10^{-4} \text{ newton/ampere-meter;}$$

the observed displacements of the ray when it hit the end of the tube were

$$d_{elec} = d_{mag} = 0.07 \text{ meters;}$$

and the lengths of the deflection and drift region were

$$\ell = 0.05 \text{ meters,} \qquad L = 1.1 \text{ meters.}$$

Inserting these values in (B.11) gives the original ray-particle velocity

$$v = \frac{(1.0 \times 10^4)\,(0.07)}{(3.6 \times 10^{-4})\,(0.07)} = 2.8 \times 10^7 \text{ meters/sec.}$$

Also, inserting these values in (B.12) gives the mass/charge ratio

$$m/e = \frac{(3.6 \times 10^{-4})^2\,(0.05)\,(1.1)\,(0.07)}{(1.0 \times 10^4)\,(0.07)^2} = 1.0 \times 10^{-11} \text{ kg/coul.}$$

This is how the values given in the last two columns of Table 2.1 were calculated.
It is also interesting to calculate the component of velocity perpendicular to the ray's original direction. Inserting (B.9) in (B.3), we see that a magnetic field B gives the ray particles a perpendicular velocity component

$$v_{\text{perp}} = eB\ell/m = B\ell/(m/e)$$

For the above stated values of B, ℓ, and m/e, this gives

$$v_{\text{perp}} = (3.6 \times 10^{-4})\ (0.05)/(1.0 \times 10^{-11}) = 1.8 \times 10^6 \text{ meters/sec}.$$

This is about 15 times smaller than the original velocity of 2.8×10^7 meters per second; so the direction and magnitude of the ray-particle velocity remained close to their original values, as was assumed in our calculation of magnetic force on the ray particles. Notice also that both v and v_{perp} are considerably less than the speed of light; so it is a good approximation to calculate the behavior of the cathode-ray particles using Newtonian mechanics, without worrying about the corrections required by Einstein's Special Theory of Relativity for particles traveling close to the speed of light.

C   Electric Fields and Field Lines

Coulomb's law states that the electric force F between two bodies that carry electric charges q_1, q_2, and are separated by a distance r has a magnitude

$$F = k_e q_1 q_2/r^2, \tag{C.1}$$

where k_e is a constant whose value depends on the system of units used to measure F, q_1, q_2, and r. With forces in newtons, charges in coulombs, and distances in meters, this constant has the value

$$k_e = 8.987 \times 10^9 \text{ newton-meter}^2/\text{coulomb}. \tag{C.2}$$

The force acts along the direction separating the two bodies. We can think of Equation (C.1) as giving the component of the force acting on either body along the direction *away* from the other body; that is, the force is repulsive if F is positive, as it is when the charges are of the same sign, and attractive if F is negative, as it is if q_1 and q_2 are of opposite sign.
It is convenient to replace (C.1) by a statement in terms of electric fields. The force on a charged body anywhere in space, say a body of charge q_1, is taken as

$$F = q_1 E, \tag{C.3}$$

where E is the electric field at the position of the body. This is to be understood as a vector equation, valid separately for each component of F and E; that is, F points in the same direction as E if q_1 is positive, and in the opposite direction if q_1 is negative. Equation (C.3) holds whatever the nature and distribution of the charges that produce the field E. For the special case of a field produced by a single body with charge q_2 at a distance r from the charge q_1, the force is given by (3.1); so the electric field must take the value

$$E = k_e q_2 / r^2 \tag{C.4}$$

pointing away from body 2 if q_2 is positive, and toward it if q_2 is negative. If the electric field is produced by several different charged bodies, then to find E we must add up (component by component) the contributions of these different bodies, each with a magnitude given by a formula like (C.4).

It is also convenient to picture the electric field in terms of field lines that pervade all space, the lines at any point having a direction taken to be the same as the direction of the electric field at that point, and the number of lines passing through a surface perpendicular to their direction taken to equal the electric field (or, if the field varies appreciably over the surface, its average) times the surface area. For instance, for the field produced by a single charged body, the electric field lines point away from the body (or toward it for negative charge); so they pass perpendicularly through any spherical surface drawn with the charged body at the center. The number of lines passing through such a spherical surface of radius r is the electric field (C.4) times the area $4\pi r^2$ of the sphere or (for a body of charge q_2)

$$\text{number of lines} = \frac{k_e q_2}{r^2} \times 4\pi r^2 = 4\pi k_e q_2. \tag{C.5}$$

Note that the radius of the sphere has canceled out; so the same number of lines pass through any spherical surface centered on q_2. Thus we can conclude that field lines do not begin or end in empty space, but are only produced and destroyed at charges, with $4\pi k_e q_2$ lines created at a positive charge q_2 and $-4\pi k_e q_2$ lines destroyed at a negative charge q_2.

The usefulness of this picture in terms of field lines is that the qualitative properties of the field lines remain unchanged even when the field is produced by many individual charges. That is, field lines do not begin or end in empty space; $4\pi k_e q$ lines leave any body with positive charge q; and $-4\pi k_e q$ lines enter any body with negative charge q. With these rules we can easily calculate the electric field in a variety of circumstances where Coulomb's law would be difficult to apply directly.

For instance, suppose we have, not a single charged point body, but a distribution of charges spread out through a sphere in some arbitrary way, the only requirement being that the distribution is spherically symmetric—that is, the distribution of electric charge is the same along any direction from the center of the sphere. The spherical symmetry of the distribution tells us that the field lines point radially outward (or inward): there is no other special direction along which we could imagine they would point. The number of lines leaving this spherical volume must be $4\pi k_e Q$, where Q is the total charge within it (for Q negative, replace "leaving" with "entering"). Hence the electric field E outside the charged sphere at a distance r from its center times the area $4\pi r^2$ of a spherical surface drawn at this distance must equal this number of lines

$$E \times 4\pi r^2 = 4\pi k_e Q$$

and therefore

$$E = \frac{k_e Q}{r^2}. \tag{C.6}$$

This may look as if we have just gone round in a circle and rederived Coulomb's law, but notice the difference: Equation (C.6) holds not just for the field produced by a charged point body at a distance r, but also for the field produced by a spherically symmetric distribution of charges spread out over some finite volume whose center is at a distance r.

To take an example of greater interest to us here, consider two flat, parallel, horizontal, metal plates on which we place electric charges of equal magnitude but opposite sign, like the plates used in Thomson's cathode-ray tube to produce an electric field that deflects the ray particles. Suppose the charges spread out uniformly over the plates. (This will actually be the case, because a nonuniform distribution would set up electric fields that would move the charges around in the conducting plates until the distribution became uniform.) Suppose also that the plates are very large compared with their separation, so that we can to a good approximation ignore the effects of their boundaries and think of them as infinite. Then the symmetry of the system tells us that the electric field lines point vertically, at right angles to the plates; there is no other special direction in the problem. Because the lines are parallel and do not begin or end between the plates, the number of lines passing at right angles through a given horizontal area is the same wherever this area is placed between the plates; so the electric field is the same everywhere between the plates. By the same reasoning, everywhere above the top plate (or below the bottom plate) the field is the same, and hence it is zero, because at a distance above the top plate sufficiently large compared with the plate separation, the field of the oppositely charged plates must cancel.

To calculate the strength of the field between the plates, we need only recall that if the plates carry charges σ and $-\sigma$ per unit area, then $4\pi k_e \sigma$ lines per unit area leave the upper plate, and they all go down into the space between the plates, because there is no field above the upper plate. The electric field between the plates is just equal to this number of lines per area

$$E = 4\pi k_e \sigma. \tag{C.7}$$

It would also be possible to obtain this result by using a formula like (C.4) to calculate the field produced by each infinitesimial bit of the upper and lower plates, and then using integral calculus to sum up these individual contributions, component by component, but it is a good deal easier to do the job by thinking about electric field lines.

D   Work and Kinetic Energy

Here we will use Newton's Second Law to derive the relation between the work done in accelerating a particle and the increase in its kinetic energy.

Suppose that a particle of mass m is accelerated by a constant force F from velocity v_1 to velocity v_2. The work W done here is the product of the force and the distance ℓ traveled by the particle:

$$W = F\ell. \tag{D.1}$$

But what is ℓ? The particle's velocity rises steadily from v_1 to v_2, so its average velocity is the average of v_1 and v_2,

$$v_{av} = \frac{1}{2}(v_1 + v_2), \tag{D.2}$$

and the distance traveled is this average velocity times the time t that the particle accelerated:

$$\ell = v_{av}t = \frac{1}{2}(v_1 + v_2)\, t. \tag{D.3}$$

But now what is t? The acceleration here is given by Newton's Second Law as F/m, and since acceleration is the change in the velocity divided by the elapsed time, we have

$$F/m = \frac{v_2 - v_1}{t},$$

or, in other words,

$$t = \frac{m(v_2 - v_1)}{F}. \tag{D.4}$$

Inserting (D.3) and then (D.4) in (D.1) now gives

$$W = F \times \frac{1}{2}(v_1 + v_2)t = F \times \frac{1}{2}(v_1 + v_2) \times m(v_2 - v_1)/F.$$

Notice that the force F cancels out. Also,

$$(v_1 + v_2)(v_2 - v_1) = v_1 v_2 - v_1{}^2 + v_2{}^2 - v_2 v_1 = v_2{}^2 - v_1{}^2;$$

so the work is

$$W = \frac{m}{2}(v_2{}^2 - v_1{}^2). \tag{D.5}$$

The kinetic energy of a particle of mass m and velocity v is defined as

$$\text{kinetic energy} \equiv \frac{1}{2}mv^2. \tag{D.6}$$

Equation (D.5) thus simply states that the increase in the particle's kinetic energy is equal to the work done on the particle.

For example, consider a particle falling in the earth's gravitational field. The force on a particle of mass m near the surface is given by Equation (A.3) as

$$F = mg, \tag{D.7}$$

where g is the acceleration due to gravity, equal to 9.8 meters/sec^2. (We will suppose that other forces, such as air resistance, are much smaller.) It follows that the work done by the earth's gravity when a particle falls from a height h_1 to a height h_2 is the force (D.7) times the distance $h_1 - h_2$ over which it acts, or

$$W = mg(h_1 - h_2). \tag{D.8}$$

Using this in (D.5), we see that the mass m can be canceled from the two sides of the equation, and we have

$$g(h_1 - h_2) = \frac{1}{2}(v_2{}^2 - v_1{}^2). \tag{D.9}$$

For instance, a weight that is dropped from rest ($v_1 = 0$) from the top of the Empire State Building ($h_1 = 300$ meters) will have a velocity v_2 when it hits the ground ($h_2 = 0$), given by

$$v_2 = \sqrt{2gh_1} = \sqrt{2 \times 9.8 \times 300} = 77 \text{ meters/second}.$$

Equation (D.9) can be rewritten in a way that shows its relation to the conservation of energy. By putting back the factor of m and moving all terms referring to the initial and final stages to the left and right sides of the equation, respectively, we have

$$\frac{1}{2}mv_1{}^2 + mgh_1 = \frac{1}{2}mv_2{}^2 + mgh_2. \tag{D.10}$$

That is, energy is conserved, provided we take account not only of the kinetic energy $\frac{1}{2}mv^2$ but also of an energy of position, or potential energy, given by

$$\text{potential energy} = mgh. \tag{D.11}$$

To see the usefulness of this interpretation in terms of the conservation of energy, consider a car rolling frictionlessly with its engine off down a mountain road. The previous derivation of (D.10) can no longer be used, because the car is under the influence of another force besides gravity: the force that the road exerts upward in resisting the weight of the car. Indeed, if the steepness of the road varies from point to point, this force is not even a constant. Nevertheless, (D.10) is still valid!—because it just states that the sum of the potential and kinetic energies is constant, which is true because there is no transfer of energy between the road and the car. (The road does exert a force on the car, but the force acts in a direction perpendicular to the surface of the road, and the car is not moving in this direction, only in the direction parallel to the road's surface.) For instance, if a car starts at rest and rolls frictionlessly down a mountain road, descending an altitude of 300 meters, then its velocity at the end of this time will be the same in magnitude (though not in direction!) as if it had fallen the same distance through empty space; that is, it will be 77 meters/second. Of course, the conservation of energy works equally well going up as down; a car that is rolling frictionlessly at a speed of 77 meters/second will be able to climb an altitude of 300 meters before it comes to rest, however steep or gentle the rise in the road.

The concept of potential energy is useful for electric as well as gravitational fields. For instance, for the charged metal plates considered in Appendix C, the electric field E is constant between the plates; so a particle of charge q will feel a constant force qE. If the top and bottom plates are positively and negatively charged, respectively, then the force on a positively charged particle is downward. By precisely the same reasoning that led to (D.11), we would here want to define an electric potential energy with the force mg replaced by qE:

$$\text{potential energy} = qEh$$

with h taken, say, as the height above the bottom plate. The voltage is the electric potential energy per charge; so the voltage at a height h above the bottom plate is given by dividing by q,

$$\text{voltage} = Eh.$$

In particular, the voltage difference between the top and bottom plates is given by setting h equal to the separation s of the two plates:

$$\text{voltage difference between plates} = Es.$$

Knowing the voltage difference produced by the electric battery to which his plates were connected, and knowing the separation s between these plates, it was easy for Thomson to calculate the electric field between the plates.

E   Energy Conservation in Cathode-Ray Experiments

Here we will show how the principle of the conservation of energy allowed Thomson and Kaufmann to calculate properties of the cathode-ray particles.

Thomson placed a collector at the end of the cathode-ray tube, and measured both the electric charge Q and the heat H deposited in it. According to the law of the conservation of energy, the heat energy must equal the total kinetic energy of the ray particles that struck the collector; if there are N of them traveling at velocity v, this gives

$$H = \frac{1}{2} mv^2 N. \tag{E.1}$$

Also, since charge is conserved, the total charge found in the collector must equal the charge of all the N cathode-ray particles that struck it:

$$Q = eN. \tag{E.2}$$

If we divide (E.1) by (E.2), the unknown N cancels out, and we have

$$H/Q = \frac{mv^2}{2e}. \tag{E.3}$$

Thomson also measured the magnetic deflection; so he knew the quantity appearing on the right-hand side of (B.10), and by dividing by the known value of $B\ell L$, he could find the quantity

$$I = mv/e. \tag{E.4}$$

If we divide (E.3) by (E.4), the unknown m/e drops out, and we have

$$v = 2H/QI. \tag{E.5}$$

This can be inserted in (E.4); solving for m/e gives, then,

$$\frac{m}{e} = \frac{I^2}{2H/Q}. \tag{E.6}$$

For instance, in the first results obtained with his "Tube 2," Thomson found the values

$$H/Q = 2.8 \times 10^3 \text{ joules/coulomb},$$
$$I = 1.75 \times 10^{-4} \text{ kg-meters/sec-coul},$$

as shown in Table 2.2 on p. 65. Then (E.5) gives

$$v = 2 \times (2.8 \times 10^3)/(1.75 \times 10^{-4}) = 3.2 \times 10^7 \text{ m/sec},$$

and (E.6) gives

$$m/e = \frac{(1.75 \times 10^{-4})^2}{2\ (2.8 \times 10^3)} = 5.5 \times 10^{-12} \text{ kg/coul,}$$

in agreement (pretty nearly) with Thomson's results in the last two columns of Table 2.2 on p. 65.

Instead of using a collector at the end of the cathode-ray tube, Kaufmann made a careful measurement of the electric potential ("voltage") difference V between the cathode and the anode of his cathode-ray tube. It was this voltage difference that was used to accelerate the cathode-ray particles to the velocity v with which, after passing through the anode, they entered the deflection region. Voltage is work per coulomb; so the work done by the electric fields in accelerating the cathode-ray particles from cathode to anode was the product of the voltage difference V and the particle charge e. But this work is also equal to the kinetic energy gained by the ray particles, and therefore

$$\frac{1}{2}\ mv^2 = eV \tag{E.7}$$

From this formula Kaufmann was able to calculate the same quantity $mv^2/2e$ that Thomson calculated using (E.3).

F   Gas Properties and Boltzmann's Constant

Here we will derive the fundamental relation between pressure, temperature, and density of dilute gases, and show how it can be used to justify Avogadro's hypothesis and to help in establishing the magnitude of the atomic scale.

The pressure of a gas is defined as the force that it exerts on any surface, per surface area. This force arises from the collisions of the gas particles with the surface. If we suppose that a gas particle striking a rigid wall exerts on it a constant force F for a time t, then according to Newton's Third Law the wall will exert on the particle a force F in the opposite direction for the same time; so the particle will be subject to an acceleration F/m (where m is the particle mass) and will suffer a change in velocity $(F/m) \times t$. If the particle gives up no energy to the wall, then its velocity after it strikes the wall must differ from its velocity before the collision only in direction, not in magnitude. Hence if the component of its velocity along the direction into the wall is $+v$ before the collision, it will be $-v$ after the collision; so the change in its velocity will be $2v$, and therefore

$$2v = Ft/m.$$

We may use this to calculate the force exerted on the wall by each particle that collides with it:

$$F = 2mv/t. \tag{F.1}$$

This formula was derived here for a particle that exerts a constant force on the wall while in contact with it, but in fact it holds even if (as will actually be true) the force varies during the collision of the particle with the wall, provided F is interpreted as the *average* force during this time. To prove this, one must break up the interval t during which the particle is in contact with the wall into tiny subintervals, each so short that, while it lasts, the force can be regarded as essentially constant. Newton's Second Law tells us that the mass times the change in the component of the particle's velocity away from the wall in each subinterval is equal to the force that the wall exerts on the particle times the duration of the subinterval. By adding up the values of both sides of this equation for all the subintervals, we see that the mass times the total change in the component of velocity away from the wall, or $2mv$, equals the sum of the durations of the subintervals, or t, times the average of the force. This summing up of infinitely many infinitesimal terms is the essential device that lies at the heart of the integral calculus.

In order to calculate the pressure, we also need to calculate the number of particles of each velocity that are in contact with the wall at any given time. This depends on the particles' velocity, as does the force (F.1) per particle. To deal with this complication, let us focus our attention on a certain area A of the wall of the vessel in which the gas is confined, and adopt the fiction that all the gas particles in the vessel have a component of velocity along the direction into this part of the wall with the same magnitude v, half of them traveling toward the wall and half away from it. We will calculate the pressure that would be exerted by the gas in this situation and then later on take account of the spread in particle velocity by averaging the pressure over all velocities.

The number of particles in contact with an area A of the wall at any time is equal to the number N of particles that strike this area in a time interval T, times the fraction t/T of the time interval that each spends in contact with the wall. The total force on this area of the wall is then the product of the force (F.1) exerted by each particle, times N, times t/T. Pressure is force per area; so the pressure here is

$$p = (2mv/t) \times N \times (t/T)/A$$

Note that the unknown time t cancels out here, and we find

$$p = 2mv \times (N/AT). \tag{F.2}$$

The quantity N/AT is just the rate at which particles hit the wall, per time and per area.

But at what rate do particles hit the wall? In a time T the particles that hit the wall will be all those that are traveling toward it and are close enough to hit it within this time, i.e., within a distance vT of the wall. The number N of particles that hit an area A of the wall in a time T is therefore equal to one-half the number in a cylinder of base area A and length vT, i.e.,

$$N = \frac{1}{2} nAvT,$$

where n is the number of these particles per unit volume in the gas. The factor $\frac{1}{2}$ arises here because, by hypothesis, half the particles are traveling toward the wall, and half away from it. We see that the rate at which particles hit the wall, per area and per time, is

$$N/AT = \frac{1}{2} nv. \tag{F.3}$$

Inserting this in (F.2), we find the pressure on the walls of the container to be

$$p = 2mv \times \frac{1}{2} nv = nmv^2.$$

As was mentioned earlier, we must average this result over the spread of particle velocities. Hence our answer for the pressure is

$$p = nm(v^2)_{av} \tag{F.4}$$

where $(v^2)_{av}$ is the average value of the square of any one component of velocity of the gas particles.

To find the value of the quantity $(v^2)_{av}$, we rely on a fundamental result of classical statistical mechanics, known as the *equipartition of energy:* Each degree of freedom of a system that has come to equilibrium will on the average have the same energy, given by

$$\bar{E} = \frac{1}{2} kT \tag{F.5}$$

where T is the temperature (measured from absolute zero), and k is the fundamental constant of statistical mechanics known as Boltzmann's constant, whose value depends on the units chosen for temperature. (The bar denotes an average over time, *not* over degrees of freedom.) It would take us too far afield to say with precision here what we mean by a "degree of freedom" of a physical system, other than that each degree of freedom furnishes one independent additive contribution to the total energy of the system. For our purposes here it is enough to note that each freely moving particle of a gas makes an additive contribution to the total energy equal to

$$\frac{1}{2} m(v_x^2 + v_y^2 + v_z^2),$$

where v_x, v_y, and v_z are the components of the particle's velocity along any three orthogonal directions, say, north, east, and up. *Each component* of each particle's velocity qualifies as an independent degree of freedom, so Equation (F.5) tells us that for a gas of freely moving particles

$$\frac{1}{2}m\overline{v_x^2} = \frac{1}{2}m\overline{v_y^2} = \frac{1}{2}m\overline{v_z^2} = \frac{1}{2}kT.$$ (F.6)

Equipartition works because, if different degrees of freedom had different average energies, then collisions or other interactions would draw energy from those degrees of freedom with higher-than-average energy and give it to the others, until their average energies were all the same. Note also that this quantity, the average energy per degree of freedom, has the essential property that we associate with temperature. If two isolated systems have different average energies per degree of freedom, and the two systems are put into contact, then energy will flow from the system with the higher energy per degree of freedom to the other, until all degrees of freedom of the combined system had the same average energy. We could, if we like, *define* the temperature of a system as the average energy in each degree of freedom, but this is not easy to measure. For historical reasons it has become nearly universal to take as the scientific unit of temperature the degree Centigrade or Celsius, defined as 1/100th of the temperature difference between the melting point of ice and the boiling point of water at normal atmospheric pressure. Boltzmann's constant provides the conversion from this everyday unit of temperature to the energy per degree of freedom. Modern measurements give its value as 1.3807×10^{-23} joules/degree. In any case, whatever units we use for temperature, Equation (F.5) gives a precise meaning to the absolute zero of temperature in classical physics: It is the temperature at which each degree of freedom has an average energy of zero. Temperature measured in degrees Centigrade but with $T = 0$ taken as absolute zero instead of the melting point of ice is said to be measured in *degrees Kelvin*, or °K, or simply K for short. On this scale, the melting point of ice is 273.16 K.

Now, back to gas pressure. Equation (F.6) gives the average over time of the square of each component of each particle's velocity. Since these are all the same, Equation (F.6) also applies if we average as well over the particles of the gas. However, now we do not need to average over time, because the conservation of energy tells us that the average over degrees of freedom of the energy per degree of freedom cannot change with time. (It equals the total energy, divided by the number of degrees of freedom.) Each component of the velocity of gas particles, averaged over particles, is thus given by

$$\frac{1}{2}m(v^2)_{av} = \frac{1}{2}kT.$$ (F.7)

The factors of $\frac{1}{2}$ cancel, and using this in (F.4), we now have

$$p = nkT.$$ (F.8)

Notice that the mass m of the gas particles has dropped out here. Thus the number of gas particles in a volume V,

$$nV = pV/kT,$$

is the same for all gases of a given volume V, pressure p, and temperature T. This is the justification for Avogadro's hypothesis.

Until a magnitude was established for the scale of atomic masses, charges, radii, etc., at the beginning of the twentieth century, physicists and chemists could not calculate the number of gas molecules in a given volume with any precision. For this reason, the gas law (F.8) was and is usually written in a rather different form. Instead of working with n, the number of gas particles per volume, one introduces the density ρ, the mass per unit volume. For gas particles of mass m, the density is

$$\rho = nm. \tag{F.9}$$

Furthermore, we can express the mass m as the molecular weight μ of the gas molecules times the mass m_1 corresponding to a molecule of unit atomic weight

$$m = \mu m_1 \tag{F.10}$$

or equivalently, since Avogadro's number N_0 is defined as $1/m_1$,

$$m = \mu/N_0. \tag{F.11}$$

Hence the gas law (F.8) can be written

$$p = \rho R T/\mu, \tag{F.12}$$

where R is the so-called *gas constant*,

$$R = k/m_1 = kN_0. \tag{F.13}$$

The point here is that by measuring the pressure, density, and temperature of gases of known molecular weight, one has a straightforward way to evaluate R. In this way, it became well-known in the nineteenth century that R has the value 8.3×10^3 joules/kg-K. With R known, a value could be found for either the Boltzmann constant k or the mass scale m_1 (or equivalently N_0) if the other were known.

For instance, in 1901, in a famous study of the thermodynamics of radiation, Max Planck was able to evaluate Boltzmann's constant as $k \simeq 1.34 \times 10^{-23}$ joules/K. Using Equation (F.13) and the value $R = 8.27 \times 10^3$ joules/kg-K of the gas constant, Planck calculated that

$$m_1 = k/R = \frac{1.34 \times 10^{-23}}{8.27 \times 10^3} = 1.62 \times 10^{-27} \text{ kg}$$

or equivalently

$$N_0 = 1/m_1 = 6.17 \times 10^{26}/\text{kg}.$$

Also, using the value of the Faraday (taken from studies of electrolysis) of

$$F \equiv e/m_1 \equiv eN_0 = 9.63 \times 10^7 \text{ coulombs/kg,}$$

Planck was able to calculate the charge of the electron

$$e = Fm_1 = 9.63 \times 10^7 \times 1.62 \times 10^{-27} = 1.56 \times 10^{-19} \text{ coulombs.}$$

A decade later, Millikan carried out a direct measurement of the electronic charge, and found

$$e = 1.592 \times 10^{-19} \text{ coulombs.}$$

Taking the value of the Faraday as $F = 9.65 \times 10^7$ coulombs/kg, Millikan was able to calculate Avogadro's number as

$$N_0 = \frac{9.65 \times 10^7}{1.592 \times 10^{-19}} = 6.062 \times 10^{26}/\text{kg,}$$

or equivalently

$$m_1 = 1/N_0 = 1.65 \times 10^{-27} \text{ kg.}$$

Also, taking the gas constant as $R = 8.32 \times 10^3$ joules/kg-K, Millikan could calculate Boltzmann's constant as

$$k = R/N_0 = \frac{8.32 \times 10^3}{6.062 \times 10^{26}} = 1.372 \times 10^{-23} \text{ joules/K.}$$

The principle of equipartition of energy also allows a simple estimate of the energy content of a gas. According to Equation (F.6), each gas particle has an average kinetic energy

$$\frac{1}{2} m\overline{v_x^2} + \frac{1}{2} m\overline{v_y^2} + \frac{1}{2} m\overline{v_z^2} = \frac{3}{2} kT.$$

If each particle has a mass m, then the energy per mass is

$$\epsilon = \frac{3}{2} kT/m = \frac{3}{2} RT/\mu.$$

This is actually only correct for monatomic gases like helium. For a gas whose molecules are diatomic, like O_2 or N_2, there are also two degrees of freedom corresponding to the two angles needed to specify the orientation of the molecule; so there is an extra energy of $2 \times \frac{1}{2} kT$ per molecule, and the energy per mass is

$$\epsilon = \frac{5}{2} RT/\mu.$$

For instance, for oxygen, $\mu = 32$; so at the typical room temperature $T = 300$ K the thermal energy in one kilogram of oxygen is

$$\frac{5}{2} \times 8.3 \times 10^3 \times 300/32 = 1.9 \times 10^5 \text{ joules}.$$

Measurements of the energy required to produce a given temperature change in a given mass of gas provide an alternative way to evaluate the gas constant R.

G   Millikan's Oil-drop Experiment

Here we will apply Newton's Second Law and the Stokes Law of viscosity to show how Millikan's measurements of the motion of oil droplets could be used to derive a value for the electric charge carried by these droplets.

Suppose that a droplet falls under the influence of gravity, with no electric field present. According to Equation (A.3) it will be subject to the downward gravity force

$$F_{\text{grav}} = mg, \tag{G.1}$$

where m is the droplet mass and $g = 9.806$ m/sec^2. The downward motion of the droplet is opposed by the viscosity of the air, which produces a force whose downward component is given by Stokes Law as

$$F_{\text{vis}} = -6\pi\eta av, \tag{G.2}$$

where $\pi = 3.14159\ldots$; η is a parameter called the viscosity of the air, which Millikan took to have the value of 1.825×10^{-5} newton-sec/m^2; a is the radius of the droplet; and v is its downward velocity. The minus sign in Equation (G.2) means that this force acts in opposite direction to the velocity, i.e., upward.

At first, when the droplet begins to fall, its velocity is small; so (G.1) is larger than (G.2), and the drop accelerates downward. Then, as the velocity rises, the magnitude of the viscous force (G.2) increases, so that the net downward force and hence the acceleration decreases. Eventually the velocity approaches a value at which (G.2) just cancels (G.1), and the droplet then falls at this velocity, with no further acceleration. We conclude, then, that the "terminal" velocity v_0 which the droplet finally reaches is found by setting the sum of Equations (G.1) and (G.2) equal to zero:

$$0 = mg - 6\pi\eta a v_0.$$ (G.3)

We know the density ρ of the droplet (its mass per volume); so we also have available a relation between m and a, that m equals the volume $4\pi a^3/3$ of the droplet times ρ:

$$m = 4\pi a^3 \rho/3.$$ (G.4)

Using (G.4) in (G.3) gives

$$0 = 4\pi a^3 \rho g/3 - 6\pi\eta a v_0$$

and we can solve for the droplet radius

$$a = \sqrt{\frac{9\eta v_0}{2g\rho}}.$$ (G.5)

Inserting this in (G.4) gives the droplet mass

$$m = \frac{4\pi\rho}{3}\left(\frac{9\eta v_0}{2g\rho}\right)^{3/2}.$$ (G.6)

Using (G.5) and (G.6), we can now deduce the mass and radius of a droplet of known density from its terminal velocity.

Now, suppose that a droplet of oil moves under the influence not only of gravity and air viscosity, but also of an electric force E pointing downward, which produces an electric force with downward component

$$F_{elec} = qE$$ (G.7)

on a droplet carrying an electric charge q. (We suppose here that q was negative, and so F_{elec} was negative, meaning that the electric force was actually in an upward direction.) With the field on, the terminal velocity reached by the oil droplet is again calculated from the condition that the acceleration and hence the total force on the droplet should vanish, but now this condition reads

$$0 = F_{grav} + F_{vis} + F_{elec}.$$ (G.8)

Using (G.1), (G.2), and (G.7), we find this to be

$$0 = mg - 6\pi\eta av + qE,$$

from which we can solve for the charge on a droplet

$$q = (- mg + 6\pi\eta av)/E. \tag{G.9}$$

Each droplet must first be observed falling, with the field off, to find m and a, and then rising, with the field on, to find q.

Before putting in numbers to see how this works, we need to mention two corrections to this simple analysis, both made by Millikan.

First, there is the buoyancy of the air. It has been understood since the time of Archimedes that the effect of buoyancy on a body immersed in a fluid is to reduce its apparent weight by an amount equal to the weight of the fluid that the body displaces. In the situations here, the buoyancy of the air reduces the effective gravity force, from the value (G.1) to

$$F_{grav} = mg - \frac{4\pi}{3} a^3 \rho_{air} g.$$

Recalling Equation (G.4), we see that the whole effect of buoyancy is just to replace the oil density ρ everywhere in our equations with an effective density,

$$\rho_{eff} = \rho - \rho_{air}. \tag{G.10}$$

The density of air at room temperature and sea-level atmospheric pressure is 1.2 kg/m^3, and the density of Millikan's oil was 0.9199×10^3 kg/m^3, so the density to be used in our equations is

$$\rho_{eff} = 0.9187 \times 10^3 \text{ kg/m}^3.$$

The second correction is rather more complicated, and also more important numerically. It arises because Stokes Law is not quite accurate for droplets so small that the droplet radius is not much larger than the average free path ℓ of air molecules between collisions. In this situation the air flowing around the droplet does not behave strictly like a smooth fluid, as Stokes had assumed, but acts to some extent like a collection of freely moving molecules. To take this into account, Millikan in effect replaced the air viscosity η with an effective viscosity, which he guessed would take the form

$$\eta_{eff} = \eta/(1 + A\ell/a), \tag{G.11}$$

where A is a constant, independent of the size of the drop or the properties of the air. A theoretical calculation had given $A = 0.788$, but Millikan found that the value $A = 0.874$ would work better, in the sense that the electronic charges measured with different droplets would come out more nearly equal if calculated using this value for A. It is this effective viscosity which must be used in Equation (G.5) to find the droplet radius a. In principle, since η_{eff} depends on a, we would then have to solve a rather complicated algebraic equation to find a. Fortunately, ℓ/a is very small; so η_{eff} is close to η, and it is therefore an adequate approximation to use the uncorrected value of a in (G.11) to find the effective viscosity

$$\eta_{eff} \simeq \eta / \left(1 + A\ell \sqrt{\frac{2g\rho_{eff}}{9\eta v_0}}\right), \tag{G.12}$$

[where we now include the buoyancy correction (G.10)], and then use this in place of η in Equation (G.5) to find the droplet radius

$$a = \sqrt{\frac{9\,\eta_{eff}v_0}{2g\rho_{eff}}} \tag{G.13}$$

and the effective droplet mass

$$m_{eff} = \frac{4\pi}{3}\rho_{eff}a^3 = \frac{4\pi}{3}\rho_{eff}\left(\frac{9\eta_{eff}v_0}{2g\rho_{eff}}\right)^{3/2}. \tag{G.14}$$

The charge on the droplet is then calculated from (G.9), using the effective values for mass and viscosity

$$q = (-m_{eff}g + 6\pi\eta_{eff}av)/E. \tag{G.15}$$

To see how this works out numerically, let us consider oil droplet number 16 in Millikan's 1911 paper. This droplet was observed to fall when the electric field was off with an average terminal speed of 5.449×10^{-4} m/sec. Taking the effective density (G.10) of the oil as 0.9187×10^3 kg/m³, the uncorrected air viscosity η as 1.825×10^{-5} newtons-sec/m², and the average free path ℓ of air molecules as 9.6×10^{-8} meters, the effective viscosity (G.12) is here

$$\eta_{eff} = \frac{1.825 \times 10^{-5}}{\left[1 + 0.874 \times 9.6 \times 10^{-8} \times \sqrt{\dfrac{2 \times 9.806 \times 0.9187 \times 10^3}{9 \times 1.825 \times 10^{-5} \times 5.449 \times 10^{-4}}}\right]}$$
$$= 1.759 \times 10^{-5} \text{ newton-sec/m}^2.$$

The droplet radius can now be calculated from Equation (G.13)

$$a = \sqrt{\frac{9 \times 1.759 \times 10^{-5} \times 5.449 \times 10^{-4}}{2 \times 9.806 \times 0.9187 \times 10^3}} = 2.188 \times 10^{-6} \text{ meters}.$$

(Millikan gave a value of 2.188×10^{-6} meters.) The effective mass of the droplet is given by Equation (G.14) as

$$m_{eff} = \frac{4\pi}{3} \times 0.9187 \times 10^3 \times (2.188 \times 10^{-6})^3$$
$$= 4.03 \times 10^{-14} \ kg.$$

With an electric field $E = 3.178 \times 10^5$ volts/meter, the droplet was observed on its first ascent to rise with a speed $v = -5.746 \times 10^{-4}$ m/sec. (The minus sign is inserted here because v was defined as the component of velocity in the downward direction, and the drop is actually rising. That is, the viscous forces here act in the same direction as gravity.) Equation (G.15) now gives the electric charge on the droplet as

$$q = [-(4.03 \times 10^{-14} \times 9.806)$$
$$- (6\pi \times 1.759 \times 10^{-5} \times 2.188 \times 10^{-6} \times 5.746 \times 10^{-4})]/3.178 \times 10^5$$
$$= -2.555 \times 10^{-18} \ coulombs.$$

This by itself does not tell us the electronic charge, because we need to know how many excess electrons were carried by the droplet. Millikan solved this problem by turning the electric field on and off a number of times, calculating the electric charge of the droplet for each of its ascents with the field on, and observing that the change in this charge between successive ascents was always close to a whole-number multiple of the same quantity of charge. Putting all this data together, Millikan in 1911 concluded that the electron has a charge $-e$ equal to $(-1.592 \pm 0.003) \times 10^{-19}$ coulombs. In particular, he could calculate that the number of electron charges carried by the droplet on its first ascent was

$$\frac{-2.555 \times 10^{-18}}{-1.592 \times 10^{-19}} = 16.05.$$

That is, droplet 16 must have carried 16 electron charges on its first ascent. The tiny discrepancy ($0.05/16 \approx 0.3$ percent) could easily be understood as due to small random errors of measurement.

The largest single error in Millikan's experiment arose not from his own measurements, but from the fact that he used a value for the viscosity of the air that we now know to be somewhat too low. The presently accepted value for η at the temperature (23°C) at which he did his experiment is 1.844×10^{-5} newton-sec/m^2, or 1 percent larger than Millikan's value. Correcting for this error has the effect of increasing η_{eff} by very close to 1 percent; increasing the droplet radius by 0.5 percent; increasing the droplet mass by 1.5 percent; and increasing all charges by 1.5 percent. In particular, after being corrected for the increase in η, Millikan's 1911 value for the electronic charge would become $(-1.616 \pm 0.003) \times 10^{-19}$ coulombs.

H   Radioactive Decay

Here we will derive the exponential law of radioactive decay, and show how it can be used to estimate the age of radioactive elements.

The half-life $t_{1/2}$ of a radioactive element is the time in which half of any sample of the element will undergo a radioactive decay. If we start with N_0 atoms of a radioactive element and wait a time t, then $t/t_{1/2}$ half-lives will have elapsed, the number of atoms will have been reduced by $t/t_{1/2}$ factors of $\frac{1}{2}$, and thus the number of remaining atoms will be

$$N = \left(\frac{1}{2}\right)^{t/t_{1/2}} N_0. \qquad \textbf{(H.1)}$$

For example, radium has a half-life of 1,600 years; so the fraction of whatever radium was present in the earth when it was formed 4.5×10^9 years ago that would still be present now is

$$\left(\frac{1}{2}\right)^{4.5 \times 10^9/1.6 \times 10^3} \simeq 10^{-850,000}.$$

The extreme smallness of this number convinces us that the radium found in the earth today must have been produced in the radioactive decay of longer-lived elements.

This sort of calculation can be turned around, to give the time required for a given decrease in radioactivity. In order to solve (H.1) for t, we must use logarithms. Recall that the logarithm of any number is the power (not necessarily a whole number) to which 10 must be raised to give that number; for instance $10^0 = 1$, $10^1 = 10$, $10^2 = 100$, etc., so

$$\log 1 = 0, \log 10 = 1, \log 100 = 2, \text{etc.}$$

Also, $10^{-1} = 0.1$, $10^{-2} = 0.01$, etc.; so

$$\log 0.1 = -1, \log 0.01 = -2, \text{etc.}$$

Further, $2 = 10^{0.3010}$, $3 = 10^{0.4771}$, etc.; so

$$\log 2 = 0.3010, \log 3 = 0.4771, \text{etc.}$$

Also, if $\log x = a$ and $\log y = b$, then $x = 10^a$ and $y = 10^b$; so $xy = 10^a \times 10^b = 10^{a+b}$ and hence

$$\log(xy) = \log x + \log y. \qquad \textbf{(H.2)}$$

Similarly

$$\log(x/y) = \log x - \log y. \qquad \textbf{(H.3)}$$

Finally, if $\log x = a$, then $x = 10^a$; so $x^y = 10^{ay}$ and hence

$$\log(x^y) = y \log x. \tag{H.4}$$

To solve Equation (H.1), we need only take the logarithm of both sides. This
gives

$$\log(N/N_0) = (t/t_{1/2}) \times \log\left(\frac{1}{2}\right) = -0.3010 \times (t/t_{1/2}). \tag{H.5}$$

For instance, the number of half-lives required for any radioactive sample to drop to 1
percent of its initial intensity is

$$t/t_{1/2} = \frac{\log(0.01)}{-0.3010} = \frac{-2}{-0.3010} = 6.64.$$

We can also use measurements of radioactive intensities to find the age of a
radioactive sample, even when only ratios of various initial abundances are known.
Suppose there are two isotopes of an element, initially produced (e.g., in stars) in the
ratio $N_{1_0}/N_{2_0} = r_0$, and now found in the ratio $N_1/N_2 = r$. Applying Equation (H.1) to
both isotopes, we have

$$N_1 = \left(\frac{1}{2}\right)^{t/t_1} N_{1_0},$$

$$N_2 = \left(\frac{1}{2}\right)^{t/t_2} N_{2_0},$$

where t_1 and t_2 are the half-lives of isotopes 1 and 2. The ratio of these two equations is

$$r = \left(\frac{1}{2}\right)^{t/t_1 - t/t_2} r_0.$$

Taking the logarithm, we find

$$\log r - \log r_0 = \left(\frac{t}{t_1} - \frac{t}{t_2}\right) \log \frac{1}{2}$$

or, solving for t,

$$t = \frac{\log r - \log r_0}{\left(\dfrac{1}{t_1} - \dfrac{1}{t_2}\right) \log \dfrac{1}{2}} \tag{H.6}$$

For instance, ^{235}U and ^{238}U have half-lives 0.714×10^9 years and 4.501×10^9 years, respectively; they were formed with an initial abundance ratio believed to be about

$$r_o \equiv (^{235}U/^{238}U)_{\text{original}} \simeq 1.65 \, ;$$

and they are now found with an abundance ratio

$$r \equiv (^{235}U/^{238}U)_{\text{now}} = 0.00723 \, .$$

Equation (H.6) then gives the age of the uranium as

$$t_U = \frac{\log (0.00723) - \log (1.65)}{\left(\dfrac{1}{0.714 \times 10^9} - \dfrac{1}{4.501 \times 10^9}\right) \times \log \dfrac{1}{2}} \, .$$

The logarithms here are

$$\begin{aligned}
\log (0.00723) &= -2.1409, \\
\log (1.65) &= 0.2175, \\
\log (1/2) &= -0.3010,
\end{aligned}$$

and we find a uranium age of 6.65×10^9 years. The universe must be at least this old.

Knowing the half-life of an element, we can calculate the rate at which individual atoms undergo radioactive decays. Suppose we start with N_0 atoms of a radioactive element and wait a very short time interval t. If N atoms are left at the end of this interval, then $N_0 - N$ atoms have decayed, and the probability that any one atom has decayed is $(N_0 - N)/N_0$. According to Equation (H.1), this is given by

$$\text{probability of decay} \atop \text{in short time } t \quad = \frac{N_0 - N}{N_0} = 1 - \left(\frac{1}{2}\right)^{t/t_{1/2}} . \qquad \text{(H.7)}$$

To evaluate this, we use the general formula for small powers of any number

$$a^\epsilon \simeq 1 + \epsilon(\log a)/M \qquad \text{(H.8)}$$

where M is the pure number $0.4343 \ldots$. Equation (H.8) is a valid approximation for values of ϵ so small that terms proportional to ϵ^2 may be neglected. Applying this to Equation (H.7) with $a = 1/2$ and $\epsilon = t/t_{1/2}$, we find that the probability of an atom decaying in a time interval t much shorter than $t_{1/2}$ is

$$\text{probability of decay} \atop \text{in a short time } t \quad \simeq - \left(\frac{t}{t_{1/2}}\right) (\log \tfrac{1}{2})/M$$

$$= \left(\frac{0.3010}{0.4343}\right)\left(\frac{t}{t_{1/2}}\right) = 0.6931 \left(\frac{t}{t_{1/2}}\right) . \qquad \text{(H.9)}$$

For instance, if we are given an atom of radium, with $t_{1/2} = 1,600$ years, then the probability that it will decay in the first 10 years we observe it is

$$0.6931 \times \frac{10}{1600} = 0.43 \text{ percent.}$$

[In order to check Equation (H.8) and to see how M is calculated, let us evaluate the $1/\epsilon$ power of the left-hand side. We can write this more neatly as

$$[1 + \epsilon(\log a)/M]^{1/\epsilon} = [(1 + \delta)^{1/\delta}]^{(\log a)/M}$$

where $\delta \equiv \epsilon (\log a)/M$. Now, for ϵ very small, δ is very small, and the quantity $(1 + \delta)^{1/\delta}$ approaches a limit called e (not to be confused with the electron charge). For instance, taking $\delta = 0.01$ or 0.0001 or 0.000001, we can calculate that

$$\begin{aligned}(1.01)^{100} &= 2.704814, \\ (1.0001)^{10000} &= 2.718146, \\ (1.000001)^{1000000} &= 2.718282.\end{aligned}$$

The convergence of these numbers demonstrates (without actually proving) that, for small δ, $(1 + \delta)^{1/\delta}$ approaches a limit close to 2.71828. A more precise value for this limit is

$$e \equiv \text{limit}_{\text{for } \delta \to 0} \text{ of } (1 + \delta)^{1/\delta} = 2.718281285.$$

Setting $(1 + \delta)^{1/\delta}$ equal to e gives

$$[1 + \epsilon(\log a)/M]^{1/\epsilon} \simeq e^{(\log a)/M} \simeq 10^{(\log e)\ (\log a)/M}.$$

We therefore take

$$M = \log e = 0.4342944819,$$

so that

$$[1 + \epsilon(\log a)/M]^{1/\epsilon} \simeq 10^{\log a} = a$$

Taking this to the power ϵ gives Equation (H.8), verifying this formula and the quoted value for M.]

The quantity $t_{1/2}/0.6931$ in (H.9) has another special significance: it is the *average life t_{av}* of each atom of the radioactive element. To see this, suppose we are given an atom of some radioactive substance, with the agreement that if it undergoes a radioactive decay, it will immediately be replaced with another atom. If we wait a time T that is very large compared with the half-life, then the number of decays we observe times the average time t_{av} between decays must equal T, and therefore

$$\text{number of decays} = T/t_{av}.$$

But since we have an atom present throughout this period, there is a uniform probability of a decay in any short time-interval t, equal to the number of decays times the fraction t/T of the total time-interval:

$$\text{probability of decay in short time } t = \frac{T}{t_{av}} \times \frac{t}{T} = \frac{t}{t_{av}}.$$

Comparing this with Equation (H.9) shows that the two formulas are consistent if and only if the average lifetime of an atom is

$$t_{av} = t_{1/2}/0.6931 = 1.4427\, t_{1/2} \tag{H.10}$$

For instance, the average lifetime of radium atoms is not equal to the half-life of 1,600 years, but to $1,600 \times 1.4427 = 2,300$ years.

　　These remarks suggest a method for using radioactivity to estimate the atomic mass m. Suppose we are able to measure the half-life of some radioactive element; for instance, we might observe that in ten years the radioactivity of a sample of radium drops to 99.568 percent of its original intensity, and using (H.5) we conclude that the half-life is

$$t_{1/2} = \frac{-0.3010 \times 10 \text{ years}}{\log (0.99568)} = 1,600 \text{ years.}$$

Suppose also that we are given a sample of known mass m of this radioactive element, small enough that we are able to count individual radioactive decays, for instance, by counting flashes of light produced when alpha particles from the radium strike a zinc-sulfide screen. The number of decays observed in a short time t will be equal to the probability (H.9) for the decay of individual atoms times the number $m/\mu m_1$ of atoms in the sample.

$$\text{decays} = 0.693 \left(\frac{t}{t_{1/2}} \right) \times \frac{m}{\mu m_1}. \tag{H.11}$$

(Here μ is the atomic weight, and m_1 is the mass corresponding to unit atomic weight; so μm_1 is the weight of one atom.) Measuring the rate of decays per time and knowing m, μ, and $t_{1/2}$, we can use this to find the atomic mass m_1, or equivalently, to find Avogadro's number $N_0 \equiv 1/m_1$.

I Potential Energy in the Atom

Here we will derive a formula for the potential energy of a charged particle at a given distance from an atomic nucleus, and use it to estimate the closest distance to the nucleus to which an alpha particle of a given velocity can penetrate.

Consider a particle with electric charge q, at a distance r from a nucleus of charge q'. The potential energy of this particle will be denoted $V(r)$, to emphasize that it is a quantity that depends on r. To find $V(r)$, imagine that the particle is pushed by the electric field of the nucleus from r to r', with r' very close to r. Since the distance traveled is very short, the force remains nearly constant during the trip, about equal to its value at r, which is given by Coulomb's law as

$$F \simeq \frac{k_e q q'}{r^2}.$$

The distance travelled is $r' - r$; so the work done by the field is $F \times (r' - r)$. But this is by definition equal to the decrease in the potential energy; so

$$V(r) - V(r') \simeq F \times (r' - r)$$

or, in other words,

$$\frac{V(r') - V(r)}{r' - r} \simeq -F = \frac{-k_e q q'}{r^2} \tag{I.1}$$

for r' very close to r. Although "$\simeq$" means "approximately equal," (I.1) should be understood as a precise statement about the behavior of $V(r') - V(r)$ as r' approaches r. In this limit, even though the numerator and denominator of the left-hand side of (I.1) both vanish, their ratio must approach a finite limit, equal to $-F$. This limit is known in calculus as the *derivative* of $V(r)$.

The condition (I.1) tells us only how $V(r)$ varies with r; it gives us no information about the value of $V(r)$ at any one value of r. Given any $V(r)$ that satisfies Equation (I.1), we can find another solution of (I.1) by simply adding a constant to $V(r)$. To fix $V(r)$, we may adopt a rather natural convention that the potential energy vanishes at very large distances from the nucleus:

$$V(r) \text{ approaches } 0 \text{ for very large } r. \tag{I.2}$$

These two conditions will suffice to determine $V(r)$.

Since work is force times distance, and the force is proportional to $1/r^2$, let us guess that V is proportional to $1/r$,

$$V(r) = A/r,$$

and use (I.1) to check this assumption and to calculate the constant A. We note that

$$V(r') - V(r) = A\left(\frac{1}{r'} - \frac{1}{r}\right) = A(r - r')/rr',$$

and therefore

$$\frac{V(r') - V(r)}{r' - r} = -A/rr'.$$

In the limit of r' approaching r, the right-hand side is $-A/r^2$; so Equation (I.1) is indeed satisfied if A is $k_e qq'$. Also (I.2) is obviously satisfied for a $V(r)$ proportional to $1/r$. We conclude then that our solution is

$$V(r) = \frac{k_e qq'}{r}. \tag{I.3}$$

It is worth stressing that (I.3) satisfies the condition (I.1) only approximately, but the approximation becomes unlimitedly precise for r' increasingly close to r; so (I.3) should be regarded as a precisely accurate solution of our problem. This sort of calculation is typical of those for which calculus was invented, and the method we have used provides one elementary example of the methods of calculus.

If an alpha particle of charge $q = 2e$ starts at infinity with energy E_∞, and has velocity v when it arrives at a distance r from a nucleus with charge $q' = Ze$, then, according to the conservation of energy, the initial energy E_∞ must equal the sum of the potential energy $V(r)$ and the kinetic energy $\frac{1}{2} mv^2$:

$$E_\infty = \frac{2k_e Ze^2}{r} + \frac{1}{2}mv^2. \tag{I.4}$$

For instance, if the alpha particle is directed straight at the nucleus, it will come to rest at a distance r_{min} given by the solution of Equation (I.4) for $v = 0$:

$$r_{min} = \frac{2k_e Ze^2}{E_\infty}. \tag{I.5}$$

If the alpha particle is artificially accelerated by a voltage difference of 10^8 volts, its energy will be 2×10^8 electron volts (because its charge is $2e$) or

$$2 \times 10^8 \times 1.6 \times 10^{-19} = 3.2 \times 10^{-11} \text{ joules.}$$

Equation (I.5) gives the distance of closest approach to the nucleus as

$$r_{\min} = \frac{2 \times 8.987 \times 10^9 \times Z \times (1.6 \times 10^{-19})^2}{3.2 \times 10^{-11}}$$

$$= 1.4 \times 10^{-17} \ Z \ \text{meters}.$$

For gold, Z equals 79; so the alpha particle penetrates to a distance of 10^{-15} meters from the center of the nucleus, close enough actually to get inside the nucleus.

J Rutherford Scattering

Here we will describe the formula derived by Rutherford for the scattering of an alpha particle by an atomic nucleus, and will show how this formula was used to verify the existence of the nucleus and to measure its charge.

Suppose an alpha particle is fired at an atom, in such a way that if it were not deflected it would miss the nucleus by a distance b. This quantity, the distance of closest approach if the forces between the alpha particle and the nucleus were miracuously turned off, is known as the *impact parameter*. By applying Newton's Second Law to the motion of the alpha particle, for each value of the impact parameter we can calculate the scatterng angle ϕ, the angle between the initial and the final directions of the alpha particle's velocity (see the figure here).

We will not be able here to go through the details of this calculation, but fortunately we can go a long way toward the final answer by a method of reasoning known as *dimensional analysis*. This method is based on the principle that the value of whatever quantity we are trying to calculate cannot depend on the units used in measuring the other quantities on which it depends. Rutherford scattering provides a nice example of the power and limitations of this method.

First, we must consider what are the input parameters on which the scattering angle ϕ might depend. It will certainly depend on the impact parameter b, and on the

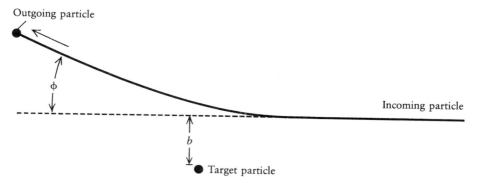

Schematic diagram of a scattering event, showing the definition of impact parameter (b) and deflection angle (ϕ).

initial speed v of the alpha particle. Also, by combining Newton's Second Law and Coulomb's Law, we see that the acceleration of the alpha particle at a distance r from the nucleus is

$$a = \frac{F}{m_\alpha} = \frac{k_e(2e)(Ze)}{m_\alpha r^2} \qquad (J.1)$$

in a direction away from the nucleus. (Recall that the alpha particle charge is $2e$, where $-e$ is the electron's charge; the nuclear charge is written Ze; m_α is the alpha-particle mass; and k_e is the constant appearing in Coulomb's Law.) Hence the scattering angle will depend on k_e, Z, e, and m_α, but *only* in the single combination*

$$2k_e Ze^2/m_\alpha. \qquad (J.2)$$

These quantities—b, v, and (J.2)—are the only input parameters on which the scattering angle ϕ can depend.

    Now, ϕ is measured in degrees or radians; so its value must be independent of the system of units used to measure distances or times or masses or charges. For instance, we know without doing any calculations at all that the correct formula for ϕ is *not* something like $\phi = 1/b$ or $\phi = 1/v$ or $\phi = 1/bv$, etc., because numerical values of these quantities do depend on the units used for lengths and times; for instance, if ϕ were equal to $1/b$, then the scattering angle would be 100 times greater if b were measured in centimeters instead of meters. So the problem is to put together b, v, and (J.2) in a *dimensionless* combination, that is, in a combination that does not depend on the units used for distances, times, etc.

    The units of (J.2) are those of acceleration times distance squared, as can be seen directly from Equation (J.1) (just bring the r^2 to the left side of the equation). Also, the units of acceleration are distance per time squared (e.g., 9.8 m/sec^2); so we can also say that the units of (J.2) are

$$2k_e Ze^2/m_\alpha \sim (\text{distance})^3/(\text{time})^2. \qquad (J.3)$$

We have no time among our input parameters, but we do have a velocity v whose units are

$$v \sim \text{distance/time}. \qquad (J.4)$$

To construct a quantity that is independent of the units used to measure time, we must therefore divide (J.2) by v^2. This yields a quantity with the units

$$2Zk_e e^2/m_\alpha v^2 \sim \text{distance}. \qquad (J.5)$$

---

* The distance r between the alpha particle and the nucleus is not included in (J.2), because it is not one of the input parameters on which ϕ might depend, but is rather a dynamical variable, and changes during the collision in a manner governed by Newton's Second Law.

Finally, to construct a quantity that is independent of the units used to measure distance *or* time, we must divide (J.5) by the only distance among our inputs, the impact parameter *b*, and obtain:

$$2Zk_ee^2/m_\alpha v^2 b. \tag{J.6}$$

Our conclusion is that the scattering angle ϕ can depend only on this one combination of input parameters. Equivalently, inverting this relation, we can say that the combination (J.6) can be expressed as some quantity $f(\phi)$ that depends only on the scattering angle:

$$2Zk_ee^2/m_\alpha v^2 b = f(\phi). \tag{J.7}$$

The impact parameter $b(\phi)$ for a given scattering angle ϕ is then given by

$$b(\phi) = 2Zk_ee^2/m_\alpha v^2 f(\phi). \tag{J.8}$$

Dimensional analysis cannot tell us anything about the nature of the quantity $f(\phi)$, but (J.8) nevertheless embodies a great deal of information about Rutherford scattering. For instance, if we are interested in scattering by some fixed angle ϕ, say, 90°, then the impact parameter is doubled if we double the nuclear charge Ze, and reduced fourfold if we double the alpha-particle velocity v. Quite a lot to learn with so little work!

Rutherford used Newtonian mechanics to calculate the orbits of the alpha particles scattered by the nucleus, and found that the impact parameter *b* and deflection angle ϕ are related by

$$b(\phi) = \frac{2Zk_ee^2}{m_\alpha v^2 \tan(\phi/2)}. \tag{J.9}$$

This has the general form (J.8) that we obtained here by dimensional analysis, and yields the further information that the quantity $f(\phi)$ has the value

$$f(\phi) = \tan(\phi/2).$$

Here "tan" is an abbreviation for the angle-dependent quantity known in trigonometry as the "tangent": if we draw a right triangle (i.e., one having a 90° or "right" angle) whose acute angles are θ and $90° - \theta$, then $\tan \theta$ is the ratio of the side of the triangle opposite the angle whose value is θ to the side opposite the angle of value $90° - \theta$. For instance, in a right triangle with both acute angles equal to 45°, the sides opposite these angles are of equal length; so their ratio is one, and therefore $\tan(45°) = 1$. Rutherford's formula (J.9) then tells us that for $\phi = 90°$, the impact parameter is

$$b(90°) = \frac{2Zk_ee^2}{m_\alpha v^2}.$$

This is just half the distance of closest approach for an alpha particle fired straight at the nucleus, calculated in Equation (I.5).

More generally, we may note that (J.9) gives the impact parameter a plausible dependence on the deflection angle. From its interpretation in terms of right triangles, it is apparent that the quantity tan θ rises steadily from a value 0 at θ = 0 to infinity at θ = 90°. It follows that b is infinite for ϕ = 0, because zero deflection is only possible when the alpha particle has missed the nucleus altogether, drops steadily with increasing ϕ, because the closer the collision, the more the deflection, and vanishes for ϕ = 180°, because the alpha particle must be fired straight at the nucleus in order to bounce straight back.

Suppose that instead of wanting to know the impact parameter for a given deflection, or vice versa, we would like to calculate the *distribution* of deflection angles for alpha particles fired at random impact parameters into a thin foil. In order to suffer a deflection *greater* than a given angle ϕ, the alpha particle must have an impact parameter of *less* than $b(\phi)$ for some atomic nucleus in the foil. We can therefore think of $b(\phi)$ as the radius of a little disc facing the stream of incoming alpha particles; an alpha particle is deflected by more than the angle ϕ if it happens to be aimed so that (if it were not deflected) it would hit one of these discs. Each disc has an effective area of π times its radius squared, or

$$\sigma = \pi b(\phi)^2, \qquad (\mathbf{J.10})$$

known as the *cross section* for scattering by at least the angle ϕ. To find out the distribution of deflection angles, we must calculate what fraction of the area of the foil is occupied by these discs. The mass M of the foil is equal to the mass m of an individual atom times the number N of atoms in the foil; so N is given by

$$N = M/m. \qquad (\mathbf{J.11})$$

Also, the mass of the foil is equal to its density ρ (mass per volume) times its volume, and the volume of the foil is given by the product of its surface area S and its thickness ℓ; so

$$M = \rho S \ell. \qquad (\mathbf{J.12})$$

Further, the mass of an individual atom can be expressed as

$$m = A/N_0, \qquad (\mathbf{J.13})$$

where A is the atomic weight and N_0 is the quantity known as Avogadro's number, defined so that $1/N_0$ is the mass of a unit atomic weight ($1/N_0 = 1.67 \times 10^{-27}$ kg). Inserting (J.12) and (J.13) into (J.11), we can write the number of atoms in the foil as

$$N = \rho S \ell N_0 / A. \qquad (\mathbf{J.14})$$

The probability $P(\phi)$ for scattering by more than the angle ϕ is given by the fraction of the total area S of the foil that is occupied by the N discs associated with the atoms of the foil, each disc with area $\sigma(\phi)$. That is, the scattering probability is

$$P(\phi) = N\sigma(\phi)/S, \qquad\qquad (\text{J}.15)$$

provided the discs do not appreciably overlap.

Inserting Equation (J.14), we see that the foil area S cancels out, and we have

$$P(\phi) = \rho\ell N_0\sigma(\phi)/A. \qquad\qquad (\text{J}.16)$$

This is a very general formula, applicable to all sorts of scattering processes. For instance, in some (but *not* all) nuclear reactions, the cross section $\sigma(0)$ for scattering by *any* angle is of the order of the geometric cross-sectional area of the nucleus, or about 2×10^{-28} m^2 for gold nuclei. Gold has a density of about 2×10^4 kg/m^3 and atomic weight 197; so the probability of scattering in gold foil is given by (J.16) as

$$(2 \times 10^4 \text{ kg/m}^3) \times \ell \times (6 \times 10^{26}/\text{kg}) \times (2 \times 10^{-28} \text{ m}^2)/197 = 12\ell,$$

where ℓ (like all the other lengths here) is expressed in meters. For relatively thick foil with $\ell = 10^{-3}$ m, the scattering probability is 1.2 percent. For thicker foils, the scattering probability approaches unity; that is, the discs begin appreciably to overlap, and the discussion above is no longer applicable.

For the special case of Rutherford scattering, the cross section $\sigma(\phi)$ is given by (J.9) and (J.10) as

$$\sigma(\phi) = 4\pi Z^2 k_e^2 e^4 / m_\alpha^2 v^4 [\tan(\phi/2)]^2. \qquad\qquad (\text{J}.17)$$

It follows that the probability (J.16) of scattering by an angle ϕ or greater is proportional to $1/[\tan \phi/2]^2$. Verification of this relation confirms that the force on the alpha particle is really proportional to the inverse square of the distance. (In particular, if the nuclear charge were spread out over a large volume, the cross section and scattering probabilities would vanish more rapidly as ϕ approaches 180°.) Also, by using (J.17) together with (J.16), we see that the scattering probability is proportional to Z^2; so measurement of this probability at any given angle allows one to find a value for the nuclear charge.

K Momentum Conservation and
Particle Collisions

Here we will describe the principle of the conservation of momentum, and use this principle to analyze the relations between particle velocities in a head-on collision.

In its most familiar form, Newton's Second Law reads

$$F = ma,$$

where F is the force acting on a particle of mass m, and a is the acceleration given to that particle. But the acceleration is the rate of change of velocity, and the mass is constant; so ma is the rate of change of the mass times the velocity v:

$$F = \text{rate of change of } mv. \qquad \textbf{(K.1)}$$

The quantity mv is known as the *momentum* of the particle. It is, like v or F, a directed quantity, being specified by its three components along three perpendicular directions, say, north, east, and up.

What is most important about momentum is the fact that it is conserved. For instance, suppose two particles called A and B come into collision. The force that B exerts on A is given by (K.1) as

$$F_{BA} = \text{rate of change of } m_A v_A.$$

and the force that A exerts on B is

$$F_{AB} = \text{rate of change of } m_B v_B.$$

But Newton's Third Law (that action equals reaction) tells us that

$$F_{BA} = -F_{AB},$$

where the minus sign indicates that the forces act in opposite directions. It follows then that

$$\text{rate of change of } m_A v_A = -\text{rate of change of } m_B v_B$$

or, in other words,

$$\text{rate of change of } m_A v_A + m_B v_B = 0. \qquad \textbf{(K.2)}$$

That is, each component of the total momentum $m_A v_A + m_B v_B$ of the two particles is *conserved:* its value after the collision is the same as before.

Now, let us apply this to a head-on collision, in which the particles recoil along the same line on which they approached each other. In this simple case, we need concern ourselves only with the components of momentum and velocity along this line. We will use subscripts 0 and 1 to distinguish velocities before and after the collision, respectively. Then (K.2) tells us that

$$m_A v_{A0} + m_B v_{B0} = m_A v_{A1} + m_B v_{B1}. \tag{K.3}$$

We have one other condition that must be imposed here: if the particles are unchanged in the collision, then the kinetic energy as well as the momentum must be conserved, so that

$$\frac{1}{2} m_A v_{A0}^2 + \frac{1}{2} m_B v_{B0}^2 = \frac{1}{2} m_A v_{A1}^2 + \frac{1}{2} m_B v_{B1}^2. \tag{K.4}$$

Normally the initial velocities v_{A0}, v_{B0} are known, and we want to calculate the final velocities v_{A1}, v_{B1}. There are two equations for these two unknowns, so a solution can generally be obtained.

To solve these equations, first solve (K.3) for v_{B1}:

$$v_{B1} = R(v_{A0} - v_{A1}) + v_{B0}, \tag{K.5}$$

where R is the mass ratio

$$R = m_A/m_B. \tag{K.6}$$

Dividing (K.4) by $m_B/2$ and inserting the value we have found for v_{B1}, we find

$$Rv_{A0}^2 + v_{B0}^2 = Rv_{A1}^2 + [R(v_{A0} - v_{A1}) + v_{B0}]^2$$
$$= Rv_{A1}^2 + R^2(v_{A0}^2 - 2v_{A0}v_{A1} + v_{A1}^2) + 2R(v_{A0} - v_{A1})v_{B0} + v_{B0}^2.$$

The v_{B0}^2 terms cancel; dividing then by R yields

$$v_{A0}^2 = v_{A1}^2 + R(v_{A0}^2 - 2v_{A0}v_{A1} + v_{A1}^2) + 2(v_{A0} - v_{A1})v_{B0}.$$

Let us neaten this by bringing together all terms with the same dependence on the unknown v_{A1}:

$$0 = (R + 1)v_{A1}^2 - 2(Rv_{A0} + v_{B0})v_{A1} + (R - 1)v_{A0}^2 + 2v_{A0}v_{B0} \tag{K.7}$$

This is a quadratic equation, and therefore has two solutions. One of these is obvious; it is easy to see that (K.7) is satisfied by $v_{A1} = v_{A0}$. I say this is obvious because it simply represents the possibility that nothing at all happens in the collision, in which case energy and momentum would of course be conserved. However, this is not the solution in which we are interested; we want to calculate the final velocities in the case where they are different from the initial ones. Nevertheless, knowing one solution of a quadratic equation is always a great help in finding the other one. Because the right-hand side of (K.7) is quadratic in v_{A1}, and vanishes when $v_{A1} = v_{A0}$, and the coefficient of the v_{A1}^2 terms is $R + 1$, it must be possible to write it in the form

$$(R + 1)v^2 - 2 (Rv_{A0} + v_{B0})\, v + (R - 1)v_{A0}^2 + 2v_{A0}v_{B0}$$
$$= (R + 1)(v - v_{A0})(v - u) \qquad \text{(K.8)}$$

The quantity v_{A1} has here been replaced with a variable v, to emphasize that this identity holds for all values of v, not just for the quantity v_{A1} that satisfies (K.7). To find u, we need only equate the values of both sides of this equation for any single value of v, say, $v = 0$. This gives

$$(R - 1)v_{A0}^2 + 2\, v_{A0}v_{B0} = (R + 1)\, v_{A0}u.$$

The value v_{A1} of v (other than v_{A0}) at which (K.8) vanishes is clearly equal to u, and therefore

$$v_{A1} = u = [(R - 1)\, v_{A0} + 2\, v_{B0}]/(R + 1).$$

Recalling the definition (K.6) of R, this result can be written more explicitly as

$$v_{A1} = [(m_A - m_B)v_{A0} + 2m_B v_{B0}]/(m_A + m_B) \qquad \text{(K.9)}$$

and inserting this back into (K.5) gives the other final velocity,

$$v_{B1} = [2m_A v_{A0} + (m_B - m_A)v_{B0}]/(m_A + m_B). \qquad \text{(K.10)}$$

The symmetry of the solution between the two particles is now apparent: v_{B1} is given by the same formula as v_{A1}, but with v_{A0} interchanged with v_{B0} and m_B interchanged with m_A.

One special case arises so often that it deserves particular mention here. If one of the particles, say, A, is initially at rest, then we must take $v_{A0} = 0$. The final velocity of the projectile particle B is then

$$v_{B1} = \left(\frac{m_B - m_A}{m_B + m_A}\right) v_{B0} \qquad \text{(K.11)}$$

and the recoil velocity of the target particle A is

$$v_{A1} = \left(\frac{2\, m_B}{m_B + m_A}\right) v_{B0}. \qquad \text{(K.12)}$$

Note that the coefficient of v_{B0} in (K.12) is always positive, unlike that in (K.11), so we find the intuitively reasonable result that the target particle can never recoil in a direction opposite to the original motion of the projectile.

These results played an important role in many of the discoveries discussed in this book. Here are a few examples.

(1) *Gas pressure.* If a particle B hits a much heavier body A, then its recoil velocity is given by (K.11) with m_A much larger than m_B. In the limit that m_B is negligible compared with m_A, this gives $v_{B1} = -v_{B0}$; that is, the projectile particle simply recoils with the same speed but reversed direction. Also, in this limit the target A recoils with what (K.12) tells us is a negligible velocity. The same results apply if A is not a particle but the wall of a chamber in which a gas of particles like B are contained: as was argued in Appendix F, the gas particles that strike the wall head-on recoil in the opposite direction with unchanged speed.

(2) *Rutherford scattering.* Geiger and Marsden observed in 1911 that alpha particles striking gold foil occasionally recoil straight backward. But (K.11) shows that a projectile particle B that strikes a stationary particle A can only recoil straight backward (i.e., v_{B1} opposite to v_{B0}) if $m_B - m_A$ is negative, i.e., if m_B is less than m_A. Rutherford could therefore conclude that the alpha particle had struck either some particle heavier than itself or some particle moving with an appreciable velocity. To deal with the second possibility, we note that, according to (K.10), a projectile particle B that collides head-on with a lighter target particle will recoil straight backward only if A was moving toward B with speed*

$$|v_{A0}| > \left(\frac{m_B - m_A}{2m_A}\right) |v_{B0}|. \tag{K.13}$$

For instance, an alpha particle has a mass 7296.3 times greater than the electron mass; so an alpha particle striking an electron head-on can recoil backward only if the electron was moving toward it with a speed more then 3647.6 times the initial speed of the alpha particle. This seemed so unlikely that one could conclude that the alpha particle must be striking some particle heavier than it is, a particle that Rutherford identified as the atomic nucleus.

(3) *Nuclear recoil in neutron scattering.* Chadwick observed that the rays produced when beryllium is exposed to alpha radiation cause the nuclei of atoms with which they collide to recoil with velocities that, for nuclei of differing atomic weights A, are proportional to the quantity

$$\frac{1}{A_0 + A}, \tag{K.14}$$

with A_0 a constant close to $A_0 = 1$. This is just what we should expect from (K.12); a projectile particle B with fixed velocity v_{B0} striking a variety of stationary target particles A will give them recoil velocities proportional to $1/(m_B + m_A)$, which is in turn proportional to (K.14) if the projectile has atomic weight A_0 and the target particles have atomic weight A. Hence from these measurements Chadwick could conclude that

---

*The vertical bars in (K.13) denote *absolute values*, that is, the magnitude of the quantities v_{A0}, v_{B0}, irrespective of their sign.

the electrically neutral particles of the beryllium rays must have an atomic weight equal to the constant A_0 in (K.14), and therefore close to one. This was correct; these particles, called neutrons, have an atomic weight 1.009.

The discussion of momentum here applies only to particles traveling at speeds much less than that of light. Einstein showed in 1905 that for particles of higher speed, the definition of momentum must be changed—but that is a subject for another book.

Physical units used in this book.

Quantity	Unit	Abbreviation	Page reference
Length	meter	m	
Time	second	sec	
Mass	kilogram	kg	26
Force	newton	N	26
Energy	joule	J	56
Electric charge	coulomb	C	37
Electric current	ampere	amp	37
Electric potential	volt	V	59
Absolute temperature	degrees Kelvin	K	184
Heat energy	calorie	cal	60

Some constants used in this book.

Quantity	Symbol	Value
Speed of light	c	2.9979246×10^8 m/sec
Electrostatic constant	k_e	8.987552×10^9 N-m /C^2
Electronic charge	e	1.60219×10^{-19} C
Electron volt	eV	1.60219×10^{-19} J
Faraday	$N_0 e$	96485 C/mole
Avogadro's number	N_0	6.0220×10^{23}/mole
Mass of unit atomic weight	m_1	1.6606×10^{-27} kg
Mass of electron	m_e	8.5473×10^{-31} kg
Mass of proton	m_p	1.67265×10^{-27} kg
Mass of neutron	m_n	1.67495×10^{-27} kg
Sidereal year	yr	3.1558×10^7 sec
Terrestrial acceleration by gravity	g	9.806 m/sec^2
Gravitational constant	G	6.672×10^{-11} N m^2/kg^2
Mechanical equivalent of heat		4.184 J/cal
Boltzmann's constant	k	1.3807×10^{-23} J/K
Circumference/diameter ratio	π	3.1415927
$(1 + \delta)^{1/\delta}$ for $\delta \to 0$	e	2.7182818

SOURCE: "Review of Particle Properties, " *Rev. Mod. Phys.* **52**, No. 2, Part II (April 1980). The uncertainty in each case is no greater than one in the last decimal place.

The elements.

Element	Symbol	Atomic number	Atomic weight
Hydrogen	H	1	1.0079
Helium	He	2	4.00260
Lithium	Li	3	6.941
Beryllium	Be	4	9.01218
Boron	B	5	10.81
Carbon	C	6	12.011
Nitrogen	N	7	14.0067
Oxygen	O	8	15.9994
Fluorine	F	9	18.998403
Neon	Ne	10	20.179
Sodium	Na	11	22.98977
Magnesium	Mg	12	24.305
Aluminum	Al	13	26.98154
Silicon	Si	14	28.0855
Phosphorus	P	15	30.97376
Sulfur	S	16	32.06
Chlorine	Cl	17	35.453
Argon	Ar	18	39.948
Potassium	K	19	39.0983
Calcium	Ca	20	40.08
Scandium	Sc	21	44.9559
Titanium	Ti	22	47.90
Vanadium	V	23	50.9415
Chromium	Cr	24	51.996
Manganese	Mn	25	54.9380
Iron	Fe	26	55.847
Cobalt	Co	27	58.9332
Nickel	Ni	28	58.70
Copper	Cu	29	63.546
Zinc	Zn	30	65.38
Gallium	Ga	31	69.72
Germanium	Ge	32	72.59
Arsenic	As	33	74.9216
Selenium	Se	34	78.96

Atomic weights here are relative to 1/12 the weight of the ^{12}C atom, and are taken from the CRC *Handbook of Chemistry and Physics,* ed. by R. C. Weast and M. J. Astle, 62nd edition (CRC Press, 1981–82)

The elements *(continued)*.

Element	Symbol	Atomic number	Atomic weight
Bromine	Br	35	79.904
Krypton	Kr	36	83.80
Rubidium	Rb	37	85.4678
Strontium	Sr	38	87.62
Yttrium	Y	39	88.9059
Zirconium	Zr	40	91.22
Niobium (Columbium)	Nb	41	92.9064
Molybdenum	Mo	42	95.94
Technetium	Tc	43	97
Ruthenium	Ru	44	101.07
Rhodium	Rh	45	102.9055
Palladium	Pd	46	106.4
Silver	Ag	47	107.868
Cadmium	Cd	48	112.41
Indium	In	49	114.82
Tin	Sn	50	118.69
Antimony	Sb	51	121.75
Tellurium	Te	52	127.60
Iodine	I	53	126.9045
Xenon	Xe	54	131.30
Cesium	Cs	55	132.9054
Barium	Ba	56	137.33
Lanthanum	La	57	138.9055
Cerium	Ce	58	140.12
Praeseodymium	Pr	59	140.9077
Neodymium	Nd	60	144.24
Promethium	Pm	61	145
Samarium	Sm	62	150.4
Europium	Eu	63	151.96
Gadolinium	Gd	64	157.25
Terbium	Tb	65	158.9254
Dysprosium	Dy	66	162.50
Holmium	Ho	67	164.9304
Erbium	Er	68	167.26
Thulium	Tm	69	168.9342
Ytterbium	Yb	70	173.04
Lutetium	Lu	71	174.967

The elements *(concluded)*.

Element	Symbol	Atomic number	Atomic weight
Hafnium	Hf	72	178.49
Tantalum	Ta	73	180.9479
Tungsten	W	74	183.85
Rhenium	Re	75	186.2
Osmium	Os	76	190.2
Iridium	Ir	77	192.22
Platinum	Pt	78	195.09
Gold	Au	79	196.9665
Mercury	Hg	80	200.59
Thallium	Tl	81	204.37
Lead	Pb	82	207.2
Bismuth	Bi	83	208.9804
Polonium	Po	84	209
Astatine	At	85	210
Radon	Rn	86	222
Francium	Fr	87	223
Radium	Ra	88	226.0254
Actinium	Ac	89	227.028
Thorium	Th	90	232.0381
Protactinium	Pa	91	231.0359
Uranium	U	92	238.029
Neptunium	Np	93	237.0482
Plutonium	Pu	94	244
Americium	Am	95	243
Curium	Cm	96	247
Berkelium	Bk	97	247
Californium	Cf	98	251
Einsteinium	Es	99	254
Fermium	Fm	100	257
Mendelevium	Md	101	257
Nobelium	No	102	259
Lawrencium	Lr	103	260

Notes for Further Reading

D. L. Anderson, *The Discovery of the Electron*. Van Nostrand, 1964.

E. N. da C. Andrade, *Rutherford and the Nature of the Atom*. Doubleday, 1964.

R. T. Beyer, ed. *Foundations of Nuclear Physics*. Dover, 1949.

J. B. Birks, ed., *Rutherford at Manchester*. Benjamin, 1965.

Sir James Chadwick, ed., *The Collected Papers of Lord Rutherford of Nelson O.M., F.R.S.* Interscience, 1963.

I. B. Cohen, "Conservation and the Concept of Electric Charge: An Aspect of Philosophy in Relation to Physics in the Nineteenth Century," in M. Clagett, ed., *Critical Problems in the History of Science*. University of Wisconsin Press, 1959.

———, *Franklin and Newton*. American Philosophical Society, 1956.

J. G. Crowther, *The Cavendish Laboratory, 1874–1974*. Science History, 1974.

B. Dibner, *Oersted and the Discovery of Electromagnetism*. Blaisdell, 1962.

A. S. Eve, *Rutherford: Being the Life and Letters of the Rt. Hon. Lord Rutherford, O.M.* Macmillan, 1939.

N. Feather, *Lord Rutherford*. Priory Press, 1973.

C. C. Gillispie, ed., *Dictionary of Scientific Biography*. Scribner's, 1970.

G. Holton, "Subelectrons, Presuppositions, and the Millikan-Ehrenhaft Dispute," in *Historical Studies in the Physical Sciences*, **9** (1978), 161.

A. J. Ihde, *The Development of Modern Chemistry*. Harper & Row, 1964.

A. I. Miller, *Albert Einstein's Special Theory of Relativity: Emergence (1905) and Early Interpretation (1905–1911)*. Addison-Wesley, 1980.

Sir Mark Oliphant, *Rutherford: Recollections of the Cambridge Days*. Elsevier, 1972.

A. Pais, "Einstein and the Quantum Theory," *Reviews of Modern Physics*, **51** (1979), 863.

———, "Radioactivity's Two Early Puzzles," in *Reviews of Modern Physics*, **49** (1977), 925.

D. Roller and D. H. D. Roller, *The Development of the Concept of Electric Charge*. Harvard University Press, 1954.

R. H. Stuewer, ed., *Nuclear Physics in Retrospect: Proceedings of a Symposium on the 1930s*. University of Minnesota Press, 1979.

George Thomson, *J. J. Thomson: Discoverer of the Electron*. Doubleday, 1965.

J. J. Thomson, *Electricity and Matter: The 1903 Silliman Lectures*. Scribner's, 1906.

———, *Recollections and Reflections*. G. Bell, 1936.

R. A. R. Tricker, *Early Electrodynamics: The First Law of Circulation*. Pergamon Press, 1965.

C. Weiner, ed., *History of Twentieth Century Physics: Course LVII of The Proceedings of the International School of Physics "Enrico Fermi."* Academic Press, 1977.

E. Whittaker, *A History of the Theories of Aether and Electricity*. Thomas Nelson, 1953.

Alexander Wood, *The Cavendish Laboratory*. Cambridge University Press, 1946.

Notes and Records of the Royal Society of London, vol. 27, no. 1, August 1972. [Articles on Rutherford by Oliphant, Massey, Feather, Blackett, Lewis, Mott, O'Shea, and Adams.]

Sources of Illustrations

frontispiece
Cavendish Laboratory.

page 5
Cavendish Laboratory.

page 13
Cavendish Laboratory.

page 17
Burndy Library.

page 21
Experimental Researches on the Electrical Discharge, by Warren de la Rue, London, 1880, Pl. 9; reprinted in *The Nineteenth Century,* L. Pearce Williams, Charles Scribner's Sons, New York, 1978.

page 23
The Science Museum, London. British Crown Copyright.

page 30
Photograph from The Science Museum, London

page 36
Burndy Library.

page 39
AIP Niels Bohr Library.

page 58
Harold Edgerton, M.I.T.

page 61
Photograph from The Science Museum, London.

page 68
Courtesy of Emilio Segrè.

page 77
Courtesy of the Archives, California Institute of Technology.

page 79
The Science Museum, London. British Crown Copyright.

page 84
Cavendish Laboratory.

page 89
By courtesy of the Royal Institution.

page 92
Photograph from The Science Museum, London.

page 95
Courtesy of the Archives, California Institute of Technology.

page 103
The Bettmann Archive, Inc.

page 104 (top)
University of Canterbury.

page 104 (bottom)
McGill University.

page 111
Ullstein Bilderdienst.

page 113
Mt. Wilson and Las Campanas Observatories, Carnegie Institution of Washington.

page 121
The Bettmann Archive, Inc.

page 133 (top)
Courtesy of the Niels Bohr Institute, Copenhagen.

page 133 (bottom)
The Museum of the History of Science, Oxford University.

page 140
Cavendish Laboratory.

page 142
AIP Niels Bohr Library.

page 143
Cavendish Laboratory.

page 148
Courtesy of the Carnegie Institution.

page 149
Courtesy of the Carnegie Institution.

page 158
C. D. Anderson, AIP Niels Bohr Library.

page 159
Courtesy of Lawrence Berkeley Laboratory.

page 161
Courtesy of Lawrence Berkeley Laboratory.

page 162
Courtesy of Lawrence Berkeley Laboratory.

page 164
Collection of E. Segrè. Courtesy of C. F. Powell.

Index

7388